Textbook of
Bakery & Confectionery

THE AUTHOR

Chef Amit Kumar Gupta Founder Secretary of Pastry Chefs and Bakers Federation of India, Masters in Tourism Management from Madurai Kamaraj University and Graduate from Delhi University. Presently holding a position Of Principal in a reputed Culinary Institute Lucknow. Chef Amit after completed his high school in 1992 started working with small bakeries as helper in Delhi and completed his apprenticeship in bakery & confectionery from National Council of Vocational Training (Taj Sats Air Catering Limited New Delhi) then serve in various hotels in bakery department. He serves several years in Carnival Cruise Lines, United States of America. Since 2011 Chef Amit serving in academics, teaching bakery and confectionery in hotel management colleges affiliated from (NCHMCT) and bakery Institute in Delhi NCR and Nagpur. He is trained trainer of baking technician, craft baker and plant baker from National Skill development Corporation India. He was chief coordinator of English Baking Camp in Hyejeon College, Hongseong County, South Korea.

Textbook of
Bakery & Confectionery

Chef Amit Kumar Gupta

Founder Secretary of Pastry Chefs &

Bakers Federation of India

2021

Daya Publishing House®

A Division of

Astral International Pvt. Ltd.

New Delhi – 110 002

Cataloging in Publication Data--DK
Courtesy: D.K. Agencies (P) Ltd. <docinfo@dkagencies.com>

Gupta, Amit Kumar (*Chef*), **author**.
Textbook of bakery & confectionery / Chef Amit Kumar Gupta.
pages cm
Includes bibliographical references.
ISBN 9789390435258 (HB)

1. Bakeries. 2. Confectionery. 3. Food service--Sanitation. 4. Food handling. I. Title. II. Title: Textbook of bakery and confectionary.
LCC TX763.G87 2021 | DDC 664.752 23

Published by : **Daya Publishing House®**
A Division of
Astral International Pvt. Ltd.
– ISO 9001:2015 Certified Company –
4736/23, Ansari Road, Darya Ganj
New DelhiA110 002
Ph. 011A43549197, 23278134
EAmail: info@astralint.com
Website: www.astralint.com

Digitally Printed at : **Replika Press Pvt. Ltd.**

Dedication

This book is devoted
to the memory of my Father
Late Prem Narain Gupta

Message

If Baking is the chemistry between a baker and his ingredients, then "Baking Science & Technology", written by Amit Gupta, is a love story. With years of experience reflecting from every single section, this book has all the essential elements required to learn baking from scratch. Though serving as a course curriculum for students, this book can come handy for budding bakers as well as experienced professional in the market. Every single detail of baking from breads and pastries to complex structures like menu designing and costing have been explained with such a detail that it makes baking a cake walk for pastry chefs and bakers. It's a great book and an amazing contribution by Chef Amit to the collection of literature on confectionary in India.

Best Regards,

Chef Arun Guwalani

Founder President

Pastry Chefs & Bakers Federation of India

Chef Patisserie Taj Group of Hotels

New Delhi

Foreword

When I try to remember my early days of culinary journey, when the passion to be a Pastry Chef was brewing in my heart. I had a lot of curiosity and zeal to learn, explore and understand some of the key fundamentals of this skill. I had many questions in my mind for which I was seeking answers and as a result, I ended up shuffling through many Pastry and Bakery books to gain the basic knowledge and understanding of this profession. Having a strong theoretical understanding is equally important as having a practical knowledge.

Today when I look at the content of this book, I'm well pleased to say that This Book has been very well curated to carter the requirement of the basics as well as the advance theoretical understanding of Pastry &Bakery Business, that can lay down a concrete foundation of Bakery Skills. Deploying the deeper understanding of every aspect of the said subject

I have known Chef Amit Gupta over a decade, and the passion that he has to share the knowledge and groom young budding chefs is commendable. I personally would like to recommend this book to the young budding chefs as well as for those Entrepreneurs, who are looking forward to setup their own business. As this can be an excellent tool to understand the commercial and culinary fundamentals of the Pastry & Bakery Skills

All the Best!

Chef. Peter D. Mello

Corporate Chef & Head New product Development

QSR/ Retail Chain of Restaurants

Foreword

I had known Chef Amit Gupta for almost 10 years now, has been my ex colleagues and a great friend. We both worked in Carnival Cruise Lines in different times.

In fact, it gives me great pleasure in sharing my thoughts about his book "Textbook of Bakery and Confectionery". This book is a result of many years of hard work and single-minded devotion of Chef Amit Gupta based on his extensive experience in Hotels and Academia.

The book scientifically designed with all the information and details required for budding professionals. Every single detail about the trade from the basic fundamentals to the advance theories which one might need to become a successful bakery professional are explained with utmost ease. I am sure that this book will serve as one of the most trusted sources of information for both beginners and seasoned chefs.

The content of the book covers all the aspects of bakery and confectionary and is designed after referring the syllabi of various Universities and Institutes. I trust it will be a great help many academicians, students and hoteliers to further enhance their knowledge.

I would recommend this book for all the students and professionals. I would like to wish all success to this book and future endeavors of Chef Amit Kumar Gupta.

Chef Vikas Singh

Program Chair, Associate Professor,
Galgotias University , Greater Noida

Preface

The aim of is Theory of Bakery and Confectionery to provide students theoretical knowledge based on National Council of Hotel Management's diploma in Bakery and Confectionery course including commodities, food costing, hygiene and sanitation.

Learning how to produce bakery and confectionery products allows the baker to produce quality products based on established procedures. This is accomplished by the study of theory and practicing skills basic to the production of food along with function of ingredients, mixing methods, food laws according to FSSAI, HACCPs, food hygiene, sanitation and cleaning procedures.

The Baker does not just produce food based on establish procedure, baker need to create and innovate. Why things are done in particular manner, how ingredients interact in the baking, the effect of temperature and a variety of other factors are the that unlock the realm of limitless possibilities in bakery and confectionery preparation which is the art and science of bakery preparation. It is the realization that for product to be attractively or even artistically presented is not sufficient. The presentation needs to be supported by desirable flavour's, textures and colours which are a result of the keys in preparation. These issues are addressed throughout the text in relation to the various types of production being discussed.

I consider bakery and confectionery preparation to be the pursuit of excellence in its most pleasing, enjoyable and satisfying form. It is hope that the National Council of Hotel Management's diploma in Bakery and Confectionery students, National Council for vocational Training apprentice, National Skill Development Corporation craft baker, baking technician and other aspiring bakers will find that the information contained in this text. Bakery and Confectionery preparation is a time honourable profession within which I wish you the brightest time- honoured and honourable profession within which I wish you the brightest and most productive of futures. Remember that humankind requires food for both the body and mind in order to grow and I encourage you to pursue both types of sustenance.

Chef Amit Kumar Gupta

Reviewed by

1. **Dr. Preeti Dixit**
Asst Professor
PSSCIVE Bhopal

2. **Nesar Ahmad Noorullah**
Executive Chef
Bhopal Baking Co.
(A unit of Jehan Numa Palace Hotel)
BBC Central Kitchen : Jehan Numa Retreat
Dr. Saleem Ali Rd. Prempura | Bhopal | Madhya Pradesh

3. **Chef Ritik Mathur**
Corporate Chef
Effingut Breweries Pvt. Ltd.
Ex- Sr. Lecturer & Acting Principal
The Lalit Suri Hospitality School,
Faridabad (NCHMCT)

4. **Rajpal Raj Bhaskar**
M.Tech (Food Process Engg.)
Agricultural & Food Engineering Department
IIT Kharagpur, West Bengal.

5. **Dinesh Sharma**
Finance Director
Carnival Cruise Line.

6 **Ranjeeta Tripathi**
Assistant Professor
Amity University

7. **Gurudutt Rohit**
Assistant Professor
Le Cordon Bleu School of Hospitality
G D Goenka University
Sohna Road, Gurugram 122103, Haryana, India

Contents

	Message (Chef Arun Guwalani)	*vii*
	Foreword (Chef. Peter D. Mello)	*ix*
	Foreword (Chef. Vikash Singh)	*xi*
	Preface	xi
	Reviewed by	*xiii*
1.	Introduction of Bakery and Confectionery	1
2.	Food Safety Hygiene and Sanitation	32
3.	Food Costing	104
4.	Bakery	130
5.	Confectionery	252
	Glossary	309
	Bibliography	337
	Index	337

Chapter 1

Introduction of Bakery and Confectionery

Source: Google image

Bakery

A bakery is a foundation that produces and sells flour-based food heated in a broiler, for example, bread, treats, cakes, baked goods, and pies. Some retail bread shops are additionally sorted as bistros, serving espresso and tea to clients who wish to expend the heated merchandise on the premises. Sweet shop things are likewise made in many bakeries all through the world.

In-store bread kitchen divisions in markets and general stores. Very good quality grocery stores, specifically, have opened numerous new open doors for innovative bread cooks and cake gourmet specialists. A couple of supermarkets have even introduced wood-consuming hearth broilers for preparing carefully assembled craftsman breads. The head cook is the expert accountable for the creation in a retail pastry kitchen. The person in question is accountable for a staff that may go from a couple of bread cooks who share most assignments to, in a bigger pastry kitchen, numerous authorities who work in various divisions.

There is no doubt that baking has always been and will continue to be one of the most important and colourful areas of culinary development. The ties of baking to agriculture lie in its quest for improved texture and refined flavours of baked goods as well as other confectioneries. The organised production of wheat by the Egyptians is considered by most historians to be the true beginning of the breads which are produced today. While it is true that the first successful bread baking is normally attributed to Asia, it was the gradual development of a better-quality domestic wheat and improved harvesting and milling methods which allowed the preparation of bread to progress in the fertile valley of the Nile

In 1765, a Parisian named Boulanger (whose name by chance, signifies "bread baker") started publicizing on his shop sign that he served soups, which he called cafés and restoratives. (Truly, the word signifies "re-establishing) According to the story, one of the dishes he served was sheep's feet in a cream sauce. The organization of stew producers tested him in court, yet Boulanger won by asserting he didn't stew the feet in the sauce yet served them with the sauce. The difficult the guidelines of the organizations. Boulanger changed the course of food administration history.

The nineteenth century saw an in-food administration as well as in the improvement of current preparing as we probably am aware it. After the French upheaval in 1789, numerous bread cooks and cake cooks who had been workers in the places of the honourability began freely business. Craftsman went after clients with the nature of their items. The overall population not simply blue-bloods and the wealthy had the option to purchase fine baked goods. A portion of the baked good shops began during that time despite everything serves Parisians today.

A creation in the eighteenth century changed the association of the business kitchen, which had been loped around an open cooking fire. This innovation was the oven, which give a more controllable warmth source. Business kitchen got isolated into three offices: the oven run by the cook, or, the rotisserie, run by the meat culinary specialist, or rôtisseur, and the broiler, run by the baked good gourmet specialist, or pâtissier. The cake culinary expert and the meat gourmet specialist answered to the Chef de Cuisine, who was otherwise called cook de food, which signify "top of the kitchen". In spite of the fact that the burner was another component of this

perceived kitchen, the pastry specialist's stove was as yet the wood-terminated block broiler that had for some time been in utilized.

The most acclaimed culinary specialist of the mid nineteenth century was Marie-Antoine Carême otherwise called Antoine Carême, who lived from 1784 to 1833. His awesome developments of sugar and baked good earned him extraordinary notoriety, and he raised the employments of cook and baked good gourmet specialist to regarded callings. Antoine Carême's book, Le pâtissier imperial, was one of the principal efficient clarifications of the baked good gourmet specialist's specialty. Incidentally, the greater part of Antoine Carême's vocation was spent in the administration of the respectability and eminence, in a period when the results of the bread cook's and cake gourmet specialist's art were getting all the more broadly accessible to average residents. Antoine Carême had little to do with the business and retail parts of preparing.

Disregarding his accomplishments and acclaim as a baked good culinary specialist, Antoine Carême was essentially not a dough puncher but rather a gourmet expert de food. As a youngster, he took in all the parts of cooking rapidly, and he committed his vocation to refining and arranging culinary strategies. His numerous books contain the main deliberate record of cooking standards, plans, and menu making.

Combination of Art and Science in Bakery and Confectionery

Bringing together a balanced quantity of various ingredients in a proper form called art and science in baking. When you mix or blend many dry ingredients, liquids and leavenings together and expose the mixture to heat, a variety of action combine as one unit. For example, butter and sugar melt, eggs coagulate, liquid makes steam. flour and starches swelling, baking powder, baking soda, and yeast form gases. Always remember whatever the condition of finish product, it is the result of the ingredients which the baker combined.

It is very important that the aspiring baker learn the performance of each ingredient used in the preparation of any product for baking. It is important to pay attention and follow the directions of a formula step by step. When something goes wrong, you must always find your mistake, or you will repeat it. The mistake must become an opportunity to learn, improve and become more professional. It is this striving to correct errors and improve your ability to reproduce a desired product that will allowed continued success for the baker.

Source: Google image

Confectionary

Confectionery is the craft of making sweets, which are food things that are wealthy in sugar and starches. When all is said in done, however, candy parlour is separated into two wide and to some degree covering classifications, dough punchers' sweets and sugar desserts.

Baker's, ice cream parlour, likewise called flour sugary treats, incorporates basically sweet baked goods, cakes, and comparable prepared products.

Sugar ice cream parlour incorporates confections (for the most part called desserts in British English), candy-coated nuts, chocolates, biting gum, bubble gum, pastillage, and different sugary treats that are made basically of sugar. At times, chocolate sugary treats (sweets made of chocolate) are treated as a different classification, as are without sugar adaptations of sugar desserts. The words treat (US and Canada), desserts (UK and Ireland), and lollies (Australia and New Zealand) are regular words for the most well-known assortments of sugar candy parlour.

Pastry specialists likewise made cakes from mixtures or players containing nectar or other sweet fixings, for example, dried natural products. A large number of these things had strict hugeness and were heated uniquely for extraordinary events, for example, Christmas cakes. Such items almost consistently had a thick surface, dissimilar to the like sweets we call cakes today. Non improved baked good batters were additionally made for such items as meat pies. During the 1400s, baked good gourmet experts in France framed their own enterprises and removed cake making from pastry specialists. Starting here on, the calling of baked good creation grew quickly, and cooks created numerous new sorts of baked good items.

Honey was the most significant sugar. Sugarcane, the wellspring of refined sugar, was local to India and developed in southern districts of Asia. To be brought to Europe, sugar needed to go through numerous nations, and each overland stop added expenses and costs to the effectively significant expense. To Europeans, sugar was an uncommon and costly extravagance thing.

The European appearance in the America in 1492 started an upheaval in baked good creation. Caribbean islands demonstrated useful for developing sugar, so the flexibly turned out to be more bountiful and costs dropped. Cocoa and chocolate, local to the New World, were accessible in the old world just because. When the new fixings turned out to be generally accessible, preparing and baked good turned out to be increasingly advanced, and numerous new plans were created. By the seventeenth and eighteenth hundreds of years, a considerable lot of the essential baked goods we know today, including covered or layered mixtures like puff baked good and Danish batter, were being made. Additionally, in the eighteenth century, processors figured out how to refine sugar from sugar beats. Europeans could develop sugar locally.

The candy parlour industry likewise incorporates specific preparing schools and broad chronicled records. Customary candy store returns to old occasions and kept on being eaten through the medieval times into the cutting edge time.

Confectionery is the art of making confections, which are food items that are rich in sugar and carbohydrates. In general, though, confectionery is divided into two broad and somewhat overlapping categories, bakers' confections and sugar confections.

SCOPE OF BAKERY AND CONFECTIONARY

Toward the start of the 21st century, fame of fine breads and cakes is becoming significantly quicker than new culinary specialists can be prepared. Those entering in a vocation in preparing or cake making today discover opportunity in numerous regions, from little bakeshops and neighbourhood cafés to enormous lodgings and discount pastry shops.

In the times past, individuals didn't make lavish dishes and adhered to basic food like chapatis, rice, beats, corn, bajra among others. There was no headway in food propensities as there was no familiarity with the propelled strategies for preparing food. Be that as it may, with the progression of time, man began thinking about some new things of food to change the taste with similar fixings utilized in customary food. Food fixings have directed the food business around the world creation individuals relish the food they eat. Food is not any more about extinguishing the yearning, yet it is really driving individuals into a fixation on the preferences it deserts. Pastry shop items are one of the most mainstream food things that individuals go for in the 21st century.

Pastry kitchen items are the readymade food things made of different fixings accessible whenever requested. The principle pastry shop items are bread, rolls, cakes, baked goods and other milk and dairy items. Individuals, because of a lack of time, incline toward these pastry kitchen items. The present-day buyer searches for new pastry kitchen items, better intrigue, taste, and comfort from bread shop nourishments.

Scope in Restaurant and Hotels Confectionary

A bakery expert was set responsible for every division. This framework, with numerous varieties, is as yet being used today, particularly in enormous inns offering conventional sorts of food administration. In a little activity, the station gourmet expert might be the main specialist in the office. In any case, in a huge kitchen, each station gourmet expert may have a few aides.

Station culinary experts in enormous kitchens incorporate the sauce gourmet specialist (saucier), who is answerable for sauces and sautéed things; the fish cook (poissonier); the meal gourmet expert (rôtisseur); and the wash room gourmet expert (culinary specialist garde trough). Sweets and baked goods are set up by the baked good culinary specialist (pâtissier). Station culinary experts report to the chief gourmet expert, or cook de food, who is responsible for food creation. In the biggest kitchens, the obligations of the leader culinary expert are for the most part administrative. The chief culinary specialist may, truth be told, do practically zero cooking by and by. The sous gourmet expert helps the leader culinary specialist and is straightforwardly accountable for the cooking during creation.

The baked good office is typically isolated truly from the hot kitchen, for at any rate two significant reasons. To start with, and generally self-evident, is that numerous treats and sweets must be set up in a cool situation. Second, the division forestalls creams, icings, and hitters from retaining the smells of cooked, barbecued, and sautéed nourishments.

In a little to medium-size eatery, the baked good culinary expert may work alone, setting up all the pastry things. Regularly, the individual in question begins work promptly toward the beginning of the day and completions before the supper administration begins. Another cook or the lounge area staff at that point collects and plates the sweets during administration.

In huge cafés and lodgings, the culinary specialist accountable for preparing and treats is the leader baked good cook. This is an administration position similar to the chief culinary expert in the hot kitchen. The chief cake culinary expert oversees laborers in the division, including authorities, for example, the bread dough puncher (Boulenger), who gets ready yeast products including such breakfast things as brioche, croissants, and Danish cake; the frozen yogurt producer (ice sheet), who makes solidified treats; the confectioner or candy creator (confiseur); and the decorator (décorateur), who plans show-stoppers, sugar work, and brightened cakes.

In inns, crafted by the heating and cake office can be broad, including planning not just treats and breads for all the on-premise eateries, bistros, and room administration, yet in addition breakfast breads and cakes and every single prepared great, including forte cakes and improving work, for the meal and providing food divisions. Such huge activities give numerous chances to the pastry specialist wishing to increase a wide scope of experience.

Food providers, institutional volume-taking care of activities (e.g., schools, medical clinics, worker break rooms), chief lounge areas, and exclusive hangouts may likewise require the administrations of cooks and baked good gourmet specialists. The necessary abilities change starting with one foundation then onto the next. Some set up the entirety of their heated merchandise in-house, while others depend on accommodation items and completed discount bread shop nourishments.

Scope in Bakery Shops

Retail bread kitchens remember autonomous bakeshops just as for store pastry shop offices in markets and general stores. Top of the line general stores, specifically, have opened numerous new open doors for inventive dough punchers and baked good gourmet experts. A couple of markets have even introduced wood-consuming hearth broilers for preparing high quality craftsman breads.

The head pastry specialist is the expert responsible for the creation in a retail bread kitchen. The individual is accountable for a staff that may extend from a couple of dough punchers who share most assignments to, in a bigger pastry kitchen, numerous masters who work in various divisions, for example, breads and yeast products, cakes, and enlivened things. Indeed, even bread-production assignments might be isolated among various specialists, with some blending, sealing, and making up the mixtures, and others preparing the things and dealing with the broilers.

Albeit most free bakeshops offer a full scope of items, from breads to cakes and baked goods, some make their notorieties on a couple of forte things, for example, cupcakes or craftsman breads, and focus on those items. More particular yet are shops whose whole business comprises of planning and designing festival cakes, for example, for weddings, birthday events, and so forth.

Discount bread kitchens achieve indistinguishable errands from retail pastry shops, however their creation offices might be more computerized and industrialized. In them, gear, for example, blenders and stoves handle huge volumes of mixtures and prepared merchandise. Notwithstanding completed things, discount bread kitchens may create incomplete items, for example, cake layers, treat mixture, and puff cake batter available to be purchased to cafés, inns, cooks, grocery stores, and other food administration task

ORGANIZATION CHART OF BAKERY

Chefs and Head Cooks

- Chefs and head prepare supervise the everyday food arrangement at eateries and different spots where food is served. They direct kitchen staff and handle any food-related concerns.

Cooks

- Cooks get ready, season, and cook a wide scope of nourishments, which may incorporate soups, plates of mixed greens, dishes, and pastries.

Food Processing Workers

- Food handling laborers work gear that blends, cooks, or procedures fixings utilized in the assembling of food and tobacco items.

Food Preparation Workers

- Food planning laborers perform numerous normal errands under the heading of cooks, culinary experts, or food administration chiefs. Food planning laborers get ready cold nourishments, cut meat, strip and cut vegetables, mix espresso or tea, and perform numerous other food administration assignments.
- Food preparation workers perform many routine tasks under the direction of cooks, chefs, or food service managers. Food preparation workers prepare cold foods, slice meat, peel and cut vegetables, brew coffee or tea, and perform many other food service tasks.

Organization Chart

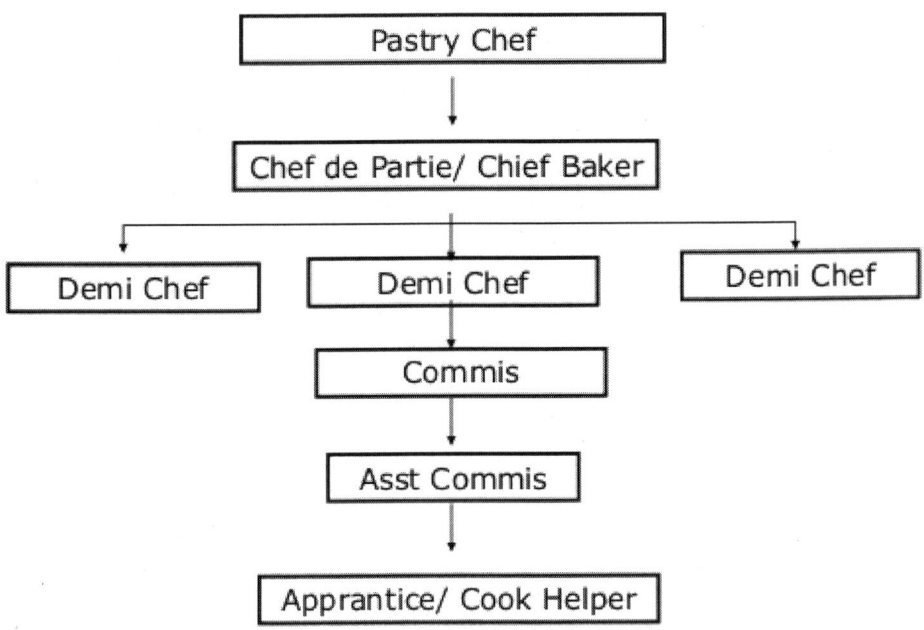

BAKERY LAYOUT

Approvals Required for Setting up Bakery

- The bakery business needs five licenses: **FSSAI license**, **GST Registration**, **Local Municipal Corporation Health License**, **Police Eating House license**, and the **Fire License**.
- Out of all the permits, FSSAI, GST, and Local Municipal Corporation Health License are the **most important** ones to have at the starting of the outlet.
- Police eating house and fire license can be acquired once you commence your operations.

- However, it is recommended to have the entire license in place before starting a bakery.

FSSAI / Food License

- You can apply online, through its website (www.fssai.gov.in).
- You can also get this done through various agencies that charge around Rs 5,000 for the complete paperwork and license fee.
- It is advised to get a five-year food license to avoid renewal fees every year.
- The fee for the five years FSSAI licenses is Rs 15,000.

GST Registration

- It can be done with the help of a Chartered Accountant.

Health License

You can get the Municipal Corporation's health license fee with the help of a local Municipal Health inspector. This would cost you around Rs 3,000.

Fire License

- You can get the fire license with a mere fee of Rs 1,000-2,000, after the installation of fire extinguisher cylinders.

Selection of Site - Location and Surroundings

- Away from environmentally polluted areas
- Industrial activities producing disagreeable or obnoxious odor, fumes, excessive soot, dust, smoke, chemical or biological emissions and pollutants, and which pose a serious threat of contaminating food
- Areas subject to flooding
- Areas prone to infestations by pests
- Areas where wastes, either solid or liquid, cannot be removed effectively
- In case, the food premise is already established and there are other industries producing pollutants, then sufficient measures such as water treatment plant, air filters etc. shall be taken to ensure food safety.
- The boundaries of the food establishment shall be clearly identified with access control. This is important to avoid the entry of stray animals.
- The external area of the food establishment should be free of debris, solid or liquid waste, stagnant water etc. to avoid pest harbourage & infestation.
- Grass/planted area inside & outside the food premises, should be regularly tended & well maintained. Preferably, insect repellent plants such as marigold, neem, basil, spearmint etc. should be grown in the near vicinity of food establishment.
- The manufacturing premise shall not have direct access to any residential area. In case residential area is in the same premises, then the entry & exit to the manufacturing premises shall be separate.

Layout Design and Construction

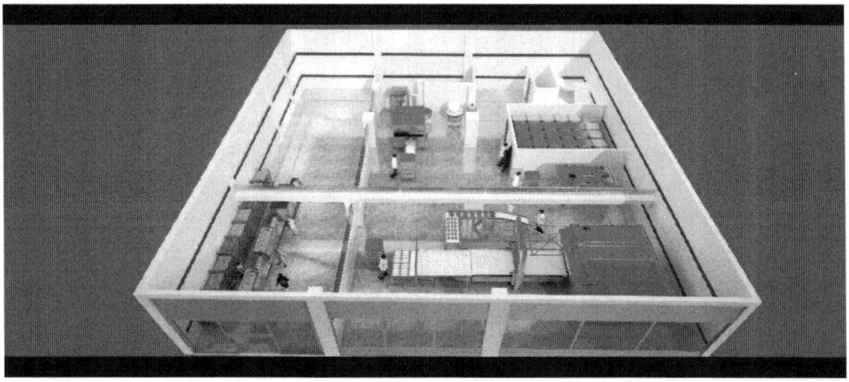

- Construction should be strong & adaptable to the weather of that area.
- Building and surrounding area shall be designed, constructed and maintained in a manner to prevent conditions, which may result in contamination of food.
- The perimeter of the building should be kept clean and tidy. It should be free of vegetation with a hard stand or gravel border where possible.
- Ledges and inaccessible areas should be kept to a minimum.
- Premises must be constructed and maintained to deny access and harbourage to pests.
- Possible entry points for pests should be proofed. Holes in the external fabric which could allow pests in should be sealed e.g. with concrete or cement and if necessary mixed with wire wool.
- All building materials shall be safe for use near food. They shall not release toxic vapours or toxic materials through contact.
- The use of materials for ceilings/roofs, which because of the environmental conditions are likely to shed particles of paint, plastic, plaster, and fibres, shall be avoided.

- External doors should be rodent proof (e.g., bristle strips, rubber compression seals for roller shutters, metal kick plates).
- External openings or doors may require proofing against insects (e.g. insect-proof screen doors, plastic strip curtains, rapid rise doors).
- Windows in other parts of the bakery and ventilation systems may require insect proofing.
- The layout must provide sufficient space to allow high-risk foods to be prepared safely.
- Either the size of the bakery and the equipment layout must allow access for cleaning around the equipment or the equipment must be moveable.
- The material movement should be done in one direction only (no backward flow), to prevent cross contamination.
- Where steam or excessive humidity occurs some form of natural or mechanical ventilation/extraction should be provided.
- Alternatively, the provision of additional heat or better insulation may prevent condensation. Where condensation is still occurring due to a particular process and/or an inherent problem with the design/construction of the building and additional ventilation is impractical the affected areas shall be cleaned and/or redecorated regularly.
- Overhead services such as cable trays, pipe runs should be kept to minimum and, if possible, a suspended ceiling installed beneath.
- Pipe and cable runs should have a smooth cleanable finish; U section fitments should be avoided or installed upside down to prevent dust and dirt collecting.
- Pipework should not be boxed in because of the risk of infestation.
- Where pipes, cables, trunking or ducting pass through internal or external walls, all holes and gaps should be filled.
- Any such external storage should be kept to a minimum, as this will provide a harbourage for pests.
- The areas should be inspected regularly, and unnecessary items removed.

Internal Structures and Fittings

- It includes floors, walls, ceiling, doors, windows, and partitions, overhead fixtures, working surface, stairs, elevators & similar structures.
- These shall be soundly constructed.
- Construction of material shall be Durable, Impervious to food particles, grease & water, Non-toxic in intended use
- Facilitate easy & effective cleaning & maintenance. Where appropriate, all junctions & corners should be rounded in processing area to facilitate effective cleaning. All the joints shall be sealed to avoid entry of pest or provide harbourage for pests.
- The structures & fittings should not act as a source of contamination such as flaking paint or plaster, seepage that may cause mould growth etc.
- Structures where glass breakage could result in contamination of food shall be constructed of alternate materials or be adequately protected.

Floor

- Floor Material - impervious, non-absorbent, washable and non-toxic materials.
- Floor surfaces should remain dry and shall be maintained in a sound condition.
- The wall-floor juncture should be rounded or coved to facilitate easy cleaning.
- The drainage system shall be properly maintained to avoid any stagnation of water and therefore do not provide any chance for pest harbourage.
- Floors shall be sloped appropriately to facilitate adequate drainage.
- The drainage shall flow in a direction opposite to the direction of food preparation / manufacturing process flow.
- The openings of the drains shall be thoroughly covered with wire mesh to prevent insects and rodents from entering the processing area.
- Floors shall be non-slippery, maintained in good repair condition and facilitate cleaning. Avoid crevices that may lead to accumulation of dirt, dust and mold growth.

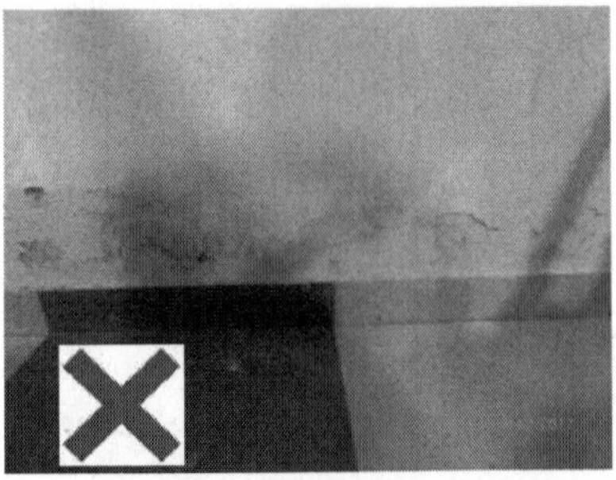

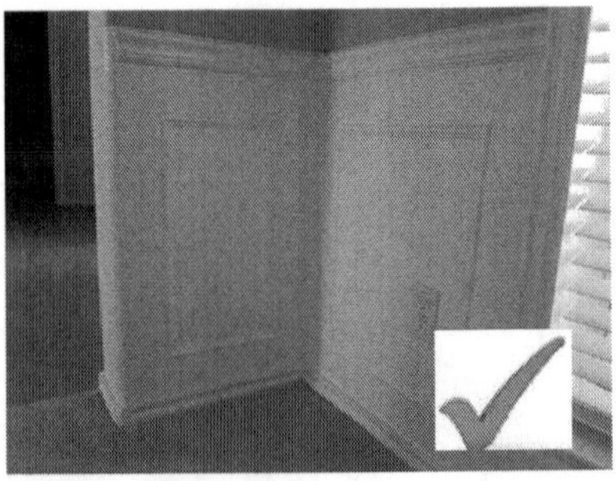

Pic from FSSAI Module

- Floor shall able to withstand operations, cleaning materials & methods. They should be strong enough to resist heavy equipment, trolley movement etc.

Wall

- Walls & partitions shall be provided to separate operations, wherever required.
- Walls shall have smooth surface, easy to clean up and with no flaking paint or plaster up to a height appropriate to operations (Height of 5 ft. is ideal) and wherever necessary, disinfect.
- Emulsion oil paint may be used which are easily cleanable by simple wiping.
- If tiles are used, they should not be porous with no crevices. Tiles also helps in avoiding seepage, thus prevents mold growth.
- SS plate or pipe may be fixed on wall to avoid damage caused to wall by moving trolleys, which causes flaking paint or plaster.
- Wall Material - impervious, non-absorbent, washable and non-toxic materials.

Ceiling

- Ceilings and overhead fixtures shall be designed, constructed, finished and maintained so as to minimize the accumulation of dirt, condensation and growth of molds and shedding of paint or plaster particles.
- Ceiling including false ceiling shall be provided where they are necessary to protect food from contamination.
- It shall be constructed to be gap free and water resistant to prevent entry of dirt, dust and pests.
- It is recommended to avoid asbestos for ceiling.
- They shall have smooth surface with no flaking paint, plaster, or loose overhead fixtures.
- All overhead fixtures such as fans, tube light etc. shall be well protected to avoid contamination of food.
- There should be no loose wires or open electrical points, as these may help in pest movement or accumulation of dirt, dust etc.

FSSAI module Pic

- Ventilators used for natural ventilation, should be easy to clean.

Window

- Adequate number of windows, roof vents or exhaust fans should be provided to ensure proper ventilation.
- Windows & other openings to the external environment shall be constructed to minimize the accumulation of dirt and shall be fitted with undamaged insect proof screens that are removable and cleanable
- Windows shall remain closed and fixed during operations.
- There should be no crevices in between walls & windows, as these could result in entry of pest.
- In case windowpanes have glass, it should be covered with tampered proof sheet.
- Wooden doors/windows should not be preferred as it promotes mold growth, termites with ageing.
- Windows, roof vents, doors & all other openings to outside environment shall be well screened with wire-mesh or insect-proof screen as applicable to protect the premise from fly and other insects / pests / animals & the doors be fitted with automatic closing springs.
- The mesh or the screen should be of a type, which can be easily removed for cleaning.
- Sufficient number of windows and exhaust openings shall be provided to minimize accumulation of dirt.

Door

- Doors shall have smooth and non-absorbent surface.
- It shall be closely fitted to avoid entry of pest, dirt, dust etc. Self-closing doors with no gap between doors and floors are preferred.
- Doors may be fitted with strip curtain, air curtain; wire mesh etc. to ensure that pest & other contaminants enter can be restricted in the food handling area.
- Door Material - smooth and non-absorbent surfaces, easy to clean and disinfect.

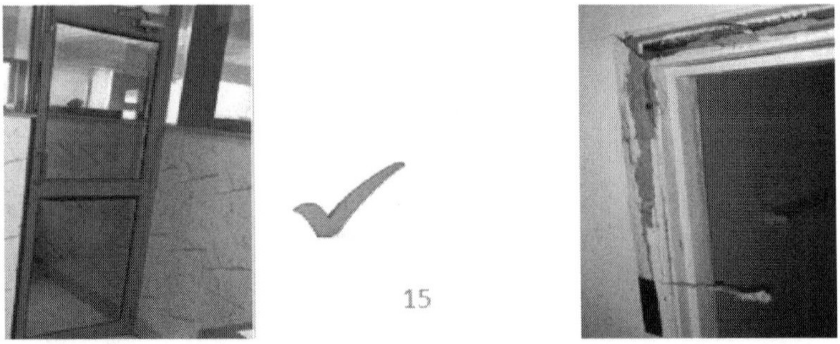

FSSAI module Pic

- Regular maintenance of door shall be conducted to prevent any mold growth or termites with ageing.

Electrical Panel

- In new plants, all socket circuits for hand-held machines must be equipped with additional protection in the form of an FI safety switch with a tripping current of 30 mA.
- The electrical panels should have rubber mats/shockproof paint flooring below to prevent from any electric shock to any employee working at the station.

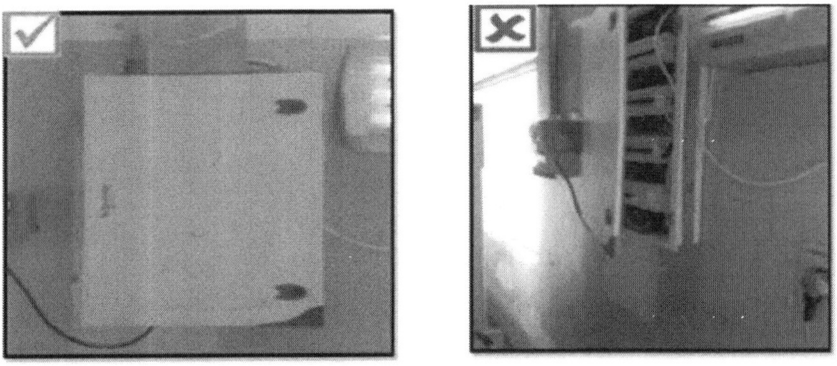

Electric panel and board should be covered

Exhaust Fan

- The exhaust fans shall be provided with flaps on outer side, the other openings shall be adequately covered with screens to avoid entry of birds and pests, and the same shall be maintained.

Light

- Adequate natural or artificial lighting should be provided to carry out operations in a hygienic manner.

- Where necessary, lighting should not be such that the resulting color is misleading. The intensity should be adequate to the nature of the operation. Uneven lighting, shadows or glare should be avoided.
- Light fittings, just above the process area, shall have shatterproof protective covers to avoid the glass, dust or insects from contaminating the food.
- Lights used in extremes of temperature e.g. in freezers or ovens, should be capable of withstanding the conditions. All walk-in chillers and freezers irrespective of size should have lights, which operate independent of the door.
- **Safety lighting** is required for the bakery and the escape routes. Storage areas and social areas should be equipped with escape route orientation lighting.
- Services, light units should be mounted flush to ceilings/ roofs or on rods or cables and preferably not chains.

- The suggested level of luminance in separate section of the bakery establishment are as follows:

S. No.	AREA	LEVEL OF LUMINANCE (lux)
1	Food storage area	220 approx
2	Inspection area	540 approx
3	General Area	220 approx
4	At food preparation surface	500 approx
5	Retail, dishwashing, hand washing, toilet areas	300 approx
6	For reading, inspection, and monitoring equipments goods are separate	600- 1200 approx
7	If the mixing space, baking room and the workplace for decorating baked	Mixing space: at least 300 Lux Baking room: at least 200 Lux Work area for decoration: at least 500 Lux.

Selection of Equipment

Material of construction

- Material of construction shall be non-corrosive, non-toxic and impervious to grease, food material, cleaning agents etc.
- The material of construction shall be in accordance with Food Safety & Standards (Packaging & Labeling) Regulation, 2011.
- It should be, durable, movable or capable of being disassembled to allow effective maintenance& cleaning.
- Chipped/enameled containers should not be used.
- Brass utensils should be provided with inner lining.
- Gasket made of food grade material should be used for food contact equipment & vessels.
- Food contact surfaces including seams should be smoothly bonded or maintained to minimize accumulation of food particles, dirt, foreign matter etc.
- Piping and ductwork shall be cleanable, drainable, and with no dead-ends.
- In case of glass containers, care should be taken for cracks & crevices. Glass containers should be limited to certain sections such as packaging. Proper cleaning should be done in case of breakage of glass in food handling area.

Maintenance and Handling

The equipment shall be such located, designed and fabricated so that they permit necessary maintenance and cleaning functions as per its intended use and facilitates good hygiene practices inside the premise including monitoring and audit.

- Equipment and utensils used in the preparation of food shall be appropriately labelled, kept at all times in good order and repair and in a clean and sanitary condition, and shall not be used for any other purpose.
- Proper Procedures should be available for handling each equipment like Ovens, Mixers, and Proofers etc.
- The equipment shall prevent the contamination of product (for example, location of lubricant reservoirs shall not be in close vicinity of the surface directly handling the food).

Cleaning

- Fixtures, fittings and equipment that use water for food handling or other activities and are designed to be connected to a water supply must be connected to an adequate supply of water.
- Equipment should be self-draining in wet process areas. Moreover, where appropriate, is connected directly to drains.
- Appropriate facilities for the cleaning and disinfecting of equipment's and instruments especially cleaning in place (CIP) system to be adopted.

Installation, Fittings, Fixtures

- It should be located /installed in a manner, which facilitates cleaning, &maintenance.
- Ideally, a distance of 1.5 ft. should be maintained in between walls & equipment & in between equipment.
- Equipment should be vented properly to prevent excessive condensation (for example, filler bowls, blanchers, retorts), where applicable
- Containers are provided with proper fitting cover/lid or with a clean gauge net or other material of texture sufficiently fine to protect the contents completely from dust, dirt etc.
- Fixtures, fittings and equipment that are designed to be connected to a sewage and wastewater disposal system and discharge sewage or wastewater must be connected to a sewage and wastewater disposal system.

Main Equipments for Food Safety

- Food Safety equipment's in bakery industry are required for keeping food products safe from foreign particles and ensuring hygienic environment all around production area.
- **Metal Detectors:** To detect metal contamination. Metal particles as small as 0.75 mm can be detected.
- **Sieves:** To sift foreign matters in flour, sugar and other ingredients.
- **Magnets:** To ensure that raw material are free from metallic properties.
- **Filters:** To prevent liquid ingredients from being contaminated.
- **Ozoniser:** Keep the environment free from microbes and other germs.
- **Air curtains:** Provides barrier to insects and dust.
- Other instruments like **Thermometer, Pressure gauge, Hygrometer** are required to monitor product and raw material quality.

Food Control and Monitoring Equipment

- The equipment used to cook, heat treat, cool, thaw, store or freeze food shall achieve the temperature as rapidly as necessary & maintain the same.
- The equipment shall also be designed to allow temperature & other characteristics (such as RH%, air flow) to be monitored & controlled, wherever required.

Container for Holding Waste, Non-Food - Chemicals & Hazardous Substance

These containers shall be

- Clearly identified for their intended purpose
- Suitably constructed of impervious material
- Easy to clean
- Leak-proof
- Provided with cover, preferable foot operated
- Located in a designated area
- No pest harborage
- Nonfood chemicals & hazardous substance shall be closed when not in use and stored separately under lock with access to only authorized personnel.
- Chemicals & other hazardous substance should be stored in original containers with label intact.

QUALITY CONTROL

Control of Operation

A. Food Receipt

1. Procurement of Raw Material
    - Raw material shall be procured from supplier having FSSAI license. Mandatory document such as Form E shall be procured.

- As a good practice, Certificate of Analysis (COA) should also be procured from the supplier mentioning its batch number, date of manufacturing, expiry date & testing parameters (physical, chemical &/or biological).
- Raw material shall meet the requirements of food product & food additives standards as laid down in regulation. As far as possible, all raw materials shall be procured in packaged form.
- Sale of loose grounded spices & cooking oil is prohibited. Raw material should be procured according to the storage & consumption capacity.
- For imported ingredients &/or products, operators should verify that the suppliers are capable of providing food products that comply with regulations laid down under Food Safety & Standard Act, 2006.

2. Vehicle Inspection
   - Vehicle should be clean & free from pest or dirt/dust etc.
   - Vehicle should be covered with tarpaulin to protect the raw material from external environment.
   - Raw material should be stacked properly.
   - Jute bags for bulk packaging are not allowed.
3. Inspection of Raw Material
   - No raw material or ingredient or any other material usedin processing products shall be accepted by a Food Business Operator, if it is known to contain chemical, physical or microbiological contaminants which would not be reduced to an acceptable level by normal sorting &/or processing.
   - All incoming material should be examined at point of receiving for physical integrity &product information mentioned on the label.
   - Receiving temperature of potentially hazardous food must be 5°C or below: or 60°C or above.
   - Receiving temperature of frozen food shall be -18°C or below.
   - Records of the receiving temperatures of potentially hazardous & frozen foods must be maintained. Where necessary, laboratory tests should be made to establish fitness for use.
   - Only sound, suitable raw materials or ingredients should be used.
   - Material should be inspected, tested or covered by COA to verify conformity with specified requirements prior to the acceptance or use.
   - The method of verification should be documented. Packaged raw material must be checked for expiry date/best before/use by date, packaging integrity & storage conditions before accepting them & stored accordingly.
   - Records of raw material or ingredient or any other material used in processing as well as their source of procurement shall be maintained for inspection & traceability.

- Access points to bulk material receiving lines should be identified, capped & locked. Discharge into such systems should take place only after approval & verification of the material to be received. Ingredients containing allergens should be clearly identified & stored to prevent cross contamination with ingredients & products not containing allergens & with other material products.
- The product specific recommended practices are as follows:

  a) Eggs

    - Eggshells should not be cracked upon receipt. Discard cracked eggs.
    - Storage of eggs in chiller until they are needed. When eggs are required to store at room temperature, current batch of eggs to be used with daily replenishment of stock.
    - Washing of hands, utensils and surfaces thoroughly with sanitizing solution and water
    - After handling eggs and before any contact with other food to prevent cross-contamination.

  b) Dry ingredients like wheat flour, sugar, milk powders, minor ingredients and cocoa powder

    - Sampling and test of incoming raw materials by appropriate test sieves.
    - Raw material should be stored in a room that has the required humidity and temperature control.

  c) Ready-to-eat products containing lightly-cooked or uncooked eggs (e.g. butter, cream, icing, mayonnaise, mousse, condensed milk)

    - Only small batches should be procured of what is required. The demand to be estimated to avoid over- production and prolonged storage.
    - Use liquid egg or egg powder instead of shell eggs where possible.
    - Cleaning of the drums containing liquid/semi-liquid raw materials for removal of debris and droppings before taking into FBO storage.
    - Storage of finished products in covered containers in the chiller at 4°C and below.
    - Storage of finished products on separate shelves above raw food (including shell eggs).

  d) Wet ingredients like Oils and Fats

    - Material carried by tanker with broken seal or without seal is not acceptable. Number on the seal is required to match with seal number mentioned in invoice / CoA. This is primarily to prevent adulteration.

- For preventing physical contamination, 30 BSS sieve and magnet to be put in the hose at the point of unloading.
- Use of proper plunger for homogenising the material in tanker or barrel before sampling.

e) Speciality Ingredients like Nuts and Dry Fruits

- Nuts received are free from fungal or insect infestation
- To ensure that adequate measures have been enforced by nut suppliers to keep equipment and processing area infestation free.
- Prepare a programme for tracking infestation levels supported by action plans for sanitation and fumigation.
- Periodic fumigation of the receiving room to prevent any possibility of cross contamination of incoming materials due to infestation Nuts are stored below 4°C to assure dormancy of the insect eggs, if any in the nuts.

4. ALLERGEN HANDLING

Major Allergens are –

a) Cereals containing gluten; i.e., wheat, rye, barley, oats, spelt or their hybridized strains and products of these;
b) Crustacean and products of these;
c) Eggs and egg products;
d) Fish and fish products;
e) Soybeans and products of these;
f) Milk and milk products (lactose included);
g) Peanut, tree nuts and nut products; and
h) Sulphite in concentrations of 10 mg/kg or more."

Allergen Control and Management

- Display all the allergens at the relevant places in the processing and storage areas for awareness among all the employees.
- All raw materials that are allergens should be labelled with a tag that states "Allergen."
- Maintain all ingredient flow during the manufacturing from non-allergen using areas to allergen using areas. This will help prevent cross-contamination.
- Preferably, products containing non-allergen ingredients should run before the product containing allergic ingredients.
- Store all allergic foods or ingredients at a designated area.

- For partially used allergic packets, the production staff should ensure the partially used packet should be stored separately and completely sealed and identified with label.
- Dedicated scoops, utensils shall be used for specific allergens.
- Thorough cleaning should be there between allergic containing product manufacture and non-allergic containing product manufacture.
- When production scheduling and cleaning operations are not performed between allergen containing production runs, allergen testing must be performed. For. E.g., ELIZA test kits are used to very.

B. STORAGE (RAW MATERIAL AND FINAL PRODUCT)

- A food business shall store food & packaging material in appropriate areas for effective protection from dust, condensation, drains, waste & other sources of contamination during storage.
- Storage areas shall be dry and well ventilated.
- All material shall be stored as per their temperature & humidity requirements& in particular section. Material shall be stored off the floor &away from walls.
- Wooden pallets are not recommended, as they are source of pest harbourage. In case wooden pallets are used, they should be fumigated once in six months.
- A distance of 0.5 feet off the floor &1.5 feet away from floor is preferred to be maintained. In case large godown are there, distance should be maintained between the pallets also, to avoid pest activity & ease of cleaning.
- The storage shall be subjected to First in First Out or First Expiry First Out approach. To meet this, adequate stock rotation systems should be in place. No expired material shall be stored in the storage section and sent to production.
- As far as possible, secondary cartons should be removed before sending the raw material/packing material in production/packaging area.
- All glass bottles should be stored at the lowermost level, to avoid contamination in case of breakage.
- Wrappers & trays should be kept under fumigation &ozonisation. All wrappers should be shrink-wrap.
- All containers used for storing raw material & finished product should be kept covered.
- Product that are sensitive to environmental conditions (for ex humidity, light) should be stored in appropriate conditions to prevent deterioration.
- During bulk flour handling &storage –
  a) House couplings, inside & outside plants, should be adequately protected from rodents, clean & in good repair.

b) Dust collectors or ventilators bags at the top of the bulk tank should be clean & insect free.

c) Inspection parts – cleanable/covered & free from contamination.

d) Tailings from sifting operations should be free from contamination.

C. FOOD PROCESSING

Major Critical Points in processing are –

S. No.	Process Step	Critical Points
1.	Issue of raw material from store to processing	• Only accepted material to be issued on First-Expiry-First-Out (FEFO) basis for raw materials and First-In-First-Out (FIFO) for packaging materials. • Take out ready-to-eat products containing lightly-cooked or uncooked eggs (e.g. butter, cream, icing, mayonnaise, mousse) what is necessary from the chiller are supplied in small batches. • Wherever thawing before use is a requirement (e.g. for compressed yeast, butter which are stored at lower temperature condition), required quantity of material should be thawed and issued at a time to Production. In case if thawed material cannot be consumed it shall be stored back in Deep Freezer. • Care to be taken to prevent deterioration of the material due to long storage under high temperature in processing area.
2.	Material Preparation	• Sieve all incoming ingredients, intermediates and add backs (if any) through appropriate standard mesh. If sieving is not feasible for example, oat flakes, viscous liquids, manual sorting and visual inspection shall be done. • Keep sieved ingredients/ additives in clean and dedicated containers / jars with proper identification, suitably above the floor. • The recommended practices while handling eggs are: • Eggs are to be washed or cleaned before use to avoid any food cross contamination. • Pool required number of eggs just before use and break them (also known as 'pooling'). • Raw eggs are to be prepared away from other food, especially cooked/ready-to-eat food to avoid cross-contamination.
3	Processing	• Use equipment namely, storage bins, sifters, dough mixers, rounder's, dough dividers, racks, slicers, proofing equipment, oven, rollers, conveyors or utensils like baking pans, pans, bowls, trays, spoons, spatulas, beaters, which are clean, free from contaminants, evidence of insect or rodent infestation and maintained in good repair. • Check the equipment for smooth edge, devoid of spot welding and any paint flaking.

S. No.	Process Step	Critical Points
		• Maintain temperature and humidity in proofing equipment, ovens and cooling area. • Maintain supply of filtered air to the processing area. • Use a fine dust mask for silo cleaning and for other heavily dust-laden activities. • Use heat protection gloves while operating ovens. • Inspection cleaning ports on flour conveyor systems shall be accessible, easy to open and clean. • Conveyor systems shall be free from loose threads and pest activity. • Working area as well as the outside premises shall be free from spilled powders, liquids, trash etc. that may attract and harbour pests, rodents and microorganisms.
4.	Premixing	• Sieve flour through minimum BSS 30 mesh supplemented with a magnetic grill. Regularly clean the sieve. • Reject consignment of flour if weevils are observed. • Introduce a periodic cleaning mechanism to prevent cross-contamination and dust generation and to ensure safe collection of unwanted materials like dust, dirt, foreign objects if any. • Use Good Practices such as vacuum cleaning, collection of debris through sodium hypochlorite. • Pass sugar through magnetic grill before use. Ensure periodic cleaning of the magnetic grill. • Use egg trays free from dirt or pests. • Collect broken eggshells in plastic bags and dispose off at regular intervals. Do not reuse such bag.
5.	Mixing	• Clean and dry mixing room without any spillage. • All mixing utensils are free from grease and old batter. • Washing of mixing bowls, beaters and scrappers with hot water at least once in 24 hours. • Place mixers, bowls and tilts above the floor level for easy access while cleaning which otherwise becomes a neglected area and a breeding place/ infestation by pests. • Use strainer for adding egg whisk while mixing. Clean the strainer with hot water at least once in each shift followed by swabbing with sodium hypochlorite solution. Keep the strainer dipped in 500-ppm sodium hypochlorite solution, when not in use. • Clean the Mixing Room floor with hot water followed by mopping with sodium hypochlorite solution or other disinfectants and floor cleaners.

S. No.	Process Step	Critical Points
6.	Forming	• Capture wet weight at pre-defined internal frequency to avoid underweight and weight variations. However, it is preferable to check the biscuit/ cookie weight at oven end rather the forming, as this creates empty band patches which lead to edge dark biscuit/ cookies, a reason for consumer non-acceptance. • Ensure thorough cleaning, verification and maintaining record wherever milk spray units or nut or sugar sprinkling units are used in forming section.
7.	Baking	• Periodically cleaning the Baking Room, followed by mopping with 500-ppm sodium hypochlorite solution or other appropriate floor odourless cleaners and maintaining it dry. • Maintain Ozoniser discharge in the baking room at 5 g per hour level, wherever necessitated. • Mop the cake cooling trolleys daily with 500-ppm sodium hypochlorite solution. • Keep the Reference samples of finished product in daylight colour cabinets with white background, to evaluate the product for its colour.
8.	Cooling	• Biscuits and bar cakes are to be transferred immediately after baking to the ambient room. • Maintaining positive pressure in ambient room. • Maintaining Ozoniser discharge in ambient room at 2 g per hour level, wherever required. • Always putting on UV lights during cooling of cakes. • Restricting entry of personnel entry in this room when the UV light is on. • Passing the product through metal detector. • Cleaning the room, mopping of floor with 500 ppm sodium hypochlorite solution or other appropriate sanitizers at least once in each shift and drying. • Checking the cooling canvas on daily basis for its physical conditions (like threads). Stitching torn canvases properly without keeping loose threads behind. • Cleaning of supporting rollers, scrapper knife and catch trays as and when required. • Use of disinfectant solution by workers for disinfecting their hands before unloading cakes. • Cakes/ Pies can also be cooled under air conditioning provided with HEPA filtered air. • Forced cooling at slab cooling room • Cleaning of the room and keeping it dry.

S. No.	Process Step	Critical Points
		• Weekly sanitizing the room with 500 ppm sodium hypochlorite solution. • Always putting on UV lights during cooling of cakes. • Restricting entry of personnel entry in this room when the UV light is on. • Maintaining room temperature at 8-100C with appropriate recording.
9.	Sandwiching/ Mallow/ Centre filled application	• Weigh raw materials for crème preparation in quantities as mentioned in the recipe. • Sieve specific ingredients of crème and/or mallow and pass through magnetic grills. • Pass sandwich pass through metal detector to avoid any presence of metal in the product.
10.	Enrobing	• Wherever pie enrobing is done, monitoring the enrobed cake weight. Passing the enrobed cake through cooling tunnel at a pre-defined temperature. • To ensure surface sanitation, then passing the enrobed cake through metal detector followed by UV light irradiation.
11.	Slicing/Packing of Products	• Cooling baked products in clean cooling chambers and tunnels. • Cooling bread by passing cool humidified air over the product. • Clearing crumbles that are left after slicing the products. • Spraying potassium sorbate uniformly on top surface of naked bar cakes before packing, if needed. • Filtering the air of sorbate spray line through an Ultra Filtration Unit periodically checked and changed. • Using clean food grade packaging to pack the products. Exposing of PVC trays, cakes and wrappers to UV light preferably before packing. • Sterilizing slicer blades and conveyor belts with isopropyl alcohol at least 3 times in each shift or as and when required. • Maintain temperature of cake slabs at the time of packing out from the slab cooling room within the range of 14 – 190C. • Keeping control samples in a separate designated place. This is required to retest the samples during any special cases like customer complaints.
12	Air Handling Unit (wherever required)	• Maintaining the Air Handling Unit inside the pre-slab and oven room. Positive pressure is maintained in the order Pre-slab room > oven room.

S. No.	Process Step	Critical Points
		• Air is blown inside the oven and Pre-slab room through sets of micro filters – first through 20 micron, then through 10 micron and finally through 5 micron filter for the oven room. • Additionally the air is passed through HEPA filter for pre-slab room. • Cleaning of 20 and 10 micron filters by water and 5 micron filter by forced air at least once in a fortnight or as required. • HEPA filter is changed when the same is choked or non-functional.

D. Food Packaging

- The packaging material shall conform to the standards as laid down under the Food Product Standards (Packaging & Labelling) Regulation, 2011.
- The packaging material shall provide all the mandatory information as required under the regulation mentioned above.
- The food packaging material shall be inspected before use to prevent using damaged, defective or contaminated packaging material, which may lead to contamination of the product.
- The packaging material & gases where used, shall be non-toxic& shall not pose threat to the safety & suitability of food under the specified conditions of storage & use.
- All the packaging surfaces should be kept clean (& sanitized for high-risk products) at all times.
- Contact parts of packaging machines to be cleaned with 500-ppm hypochlorite solution.PVC trays, cakes & wrappers should be exposed to UV light during packing.
- Food handlers should use sterilized gloves for manually handling naked cakes or disinfect their hands with disinfectant solution.
- Wrapping and packaging operations shall be carried out in a hygienic manner so as to avoid contamination of the products. First In First Out or First Expiry First Out approach shall be followed. To meet this, adequate stock rotation to be followed. Packing room should be air conditioned in case of cake or pie (preferably maintained at 22 – 30°C).
- Use of staple pins, strings, rubber bands should be avoided. Glue, if used, should not come in contact with the food product, & in case it comes in contact, it shall be food grade.
- Reusable packaging, if used, shall be suitably durable, easy to clean & where necessary, disinfect. It shall not have been used for packaging non – food products.

E. Rework & Control of Non-Conforming Products

- A non-conforming product can be detected through –

  a) Customer complaints

b) Internal defect findings
c) Internal audits
d) External audits
e) Incoming material inspections
f) Regular testing & inspection activities.
    - These products should either be disposed off or reworked.
    - On conforming that product is non-conforming, the product shall be clearly identified, kept labelled & segregated to allow traceability. All traceability records of rework shall be maintained such as product name, production date, batch no etc.
    - Stored material for rework shall be protected from exposure to microbiological, chemical or extraneous matter contamination.
    - Where rework is incorporated into a product as an "in-process" step, the acceptable quantity, the process step & method of addition, including any necessary pre-processing stages, shall be defined.
    - Where rework activities involve removing a product from filled or wrapped packages, control shall be put in place to ensure the removal & segregation of packaging materials & to avoid contamination of the product with extraneous matter.
    - Handling of Allergen rework/ add back to be done in such a way, that the rework containing allergen shall not cross contaminate non allergen containing food material during processing, handling & storage.
    - Non-conforming material to be disposed off shall be disposed as per the requirements of Environment Protection Act, 1986.

F. Food Transportation

- Conveyances &/or containers used for transporting food stuffs shall be designed and constructed to permit adequate cleaning &/or disinfection.
- They shall be kept clean & maintained in good repair condition to protect food stuff from contamination.
- Identification of an appropriate supply chain shall be done having the provision of minimizing food spoilage during transportation.
- Use non toxic materials of construction for conveyances meant for transporting food stuff Where direct contact with food may occur, materials used in carrier construction should be suitable for food contact.
- Inspection of vehicle prior to loading with foods for any visual evidence of contamination and/or pest infestation, dirt or other obnoxious matter and maintaining records thereof.
- Applying FEFO for dispatch of all products. Loading of goods should be done in designated area with sufficient ventilation, preferably in separate

rooms. If possible practice controls like conveyorization, usage of loading stands to avoid product breakage/ integrity while loading.

- Ideally vehicles should not be used for transporting food stuff other than food stuffs where this may result in contamination of foodstuff. However, where conveyances are used for transportation anything other than food stuffs or for transporting different foods, there shall be effective cleaning between loads to avoid risk of contamination.
- Wherever necessary, conveyances and/or containers are used for transporting anything in addition to foodstuffs or for transporting different foods at the same time, there shall be, where necessary, effective separation of products to prevent cross contamination.
- Wherever necessary, conveyances used for transporting food stuff shall be capable of maintaining food stuffs at appropriate temperatures & allow those temperatures to be monitored and maintaining records thereof.

G. Food Traceability & Recall

- Effective recall procedure where applicable, including a description of the responsibilities & actions to be taken, shall be in place to deal with any food safety hazard & to enable the complete, rapid recall of any implicated lot/batch of the finished food from the market.
- For recall to be performed, Food Safety & Standard (Food Recall Procedure) Regulation, 2011 shall be followed.
- The recall program should be tested at least once a year through appropriate means such as mock recall. It should be tested to verify its ability to rapidly identify, control & recall all the potentially affected products.
- The operator should identify & correct any deficiencies in the recall procedure. Recalled products shall be held under supervision until they are destroyed, used for purposes other than human consumption, determined to be safe for human consumption, or reprocessed in a manner to ensure food safety.
- Records of recalled products shall be maintained. Records should demonstrate effective identification, traceability /product tracing, recall, subsequent handling & disposal of recalled food.

H. Quality Control

- The food business operator shall have a quality control programme in place to include inspection and testing of incoming, in process and finished products.
- Calibration of laboratory equipment's shall be done periodically.
- Adequate infrastructure including the laboratory facility and trained and competent testing personnel should be available for carrying out testing.
- In case of inadequate in house test facilities, a system shall be in place for testing these materials in an accredited external laboratory/ laboratory notified by FSSAI.

- In case of complaints, the food business operator shall carry out testing either through their in-house/external NABL accredited laboratories notified by FSSAI.
- Finished food product shall be tested as per FSS standards and regulations 2011 at least once in six months from an FSSAI notified laboratory.
- It is recommended to retain the control samples, till the end of the shelf life. Testing records shall be maintained.

Chapter 2

Food Safety Hygiene and Sanitation

FOOD MICROBIOLOGY

1. Introduction

Food science is a discipline concerned with all aspects of food - beginning after harvesting, and ending with consumption by the consumer. It is considered one of the agricultural sciences, and it is a field, which is entirely distinct from the field of nutrition.

The field of food microbiology is a very broad one, encompassing the study of microorganisms, which have both beneficial and deleterious effects on the quality, and safety of raw and processed foods. It is important to understand the relationships among the various microorganisms making up the microflora of a food. In fact, food microbiologists are concerned with the practical implications of the microflora of the food and the food microorganisms that can cause spoilage of food and disease in humans. Of course, some useful bacteria may be important because they change the functional properties of foodstuffs resulting in new tastes, odors or textures.

Microorganisms in food include bacteria, molds, yeasts, algae, viruses, parasitic worms and protozoa. These organisms differ in size and shape and in their biochemical and cultural characteristics.

2. Microorganism Groups Important in Food Microbiology

Viruses

- Viruses are 10- 450 nm in size; cannot reproduce without a living host; attack only susceptible host cell lines; infect plants, animals, and bacteria; and have the capacity to produce specific diseases in specific hosts.
- Transmission occurs in foods, water and air.
- Viruses are too small to be visualized with an ordinary compound microscope. Only after the electron microscope was developed, the direct observation of viruses was possible.
- Viruses consist of a DNA or RNA core surrounded by a protein coat. Because they lack all the apparatus for normal cellular metabolism, they must utilize the cellular machinery of the host cell in order to grow and divide. Once they invade a host cell, however, viruses can multiply very rapidly.

- Viruses are important in food for three reasons. Some are able to cause enteric disease and thus, if present in a food, can cause food borne diseases.
- Hepatitis A and Norwalk viruses have been implicated in food borne outbreaks.
- Several other enteric viruses, such as Poliovirus, Echovirus, and Coxsackievirus, have the potential of causing food borne diseases.
- In some countries where the level of sanitation is not very high, they can contaminate foods and cause disease.
- Some bacterial viruses (bacteriophages) are used in the identification of species/ strains by a process called tranduction (e.g., in Escherichia, coli, Lactococcus lactis).
- Finally, some bacteriophages can be very important due to their ability to cause fermentation failure.
- Many lactic acid bacteria, used as starter cultures in food fermentation, are sensitive to different bacteriophages.
- These phages can infect and destroy starter culture bacteria, causing product failure.
- Methods are being studied to genetically engineer lactic acid start cultures so that they become resistant to multiple bacteriophages.

Table: Human Intestinal Viruses with High Potential as Food Contaminants

Types of Viruses	Example
Picornaviruses	Polioviruses Coxsackievirus A Coxsackievirus B Echovirus Enterovirus
Reoviruses	Reovirus Rotavirus
Parvoviruses	Human gastrointestional viruses
Papovaviruses	Human BK and JC viruses
Adenoviruses	Human adenoviruses

Bacteria

- Bacteria are unicellular Micro-organisms that are approximately one micro meter (10-3mm) in diameter with variations in morphology from short and elongated rods (bacilli), spherical or ovoid forms(cocci), vibrio (comma shaped) and even spiral in shape.
- Bacteria produce various pigments, which range from shades of yellow to dark pigments such as brown or black.
- Certain bacteria have pigmentation of intermediate colors such as red, pink, orange, blue, green, or purple.

- These bacteria cause food discoloration, especially, among foods with unstable color pigments such as meat.
- Some bacteria also cause discoloration by slime formation.
- Among the Microorganisms found in foods, bacteria constitute a major important group.
- This is not only because many different species can be present in foods, but is also due to their rapid growth rate, ability to utilize food nutrients, and their ability to grow under a wide range of temperatures, aerobiosis, pH, and water activity, as well as to survive under adverse situations, such as survival of spores at high temperature.
- For convenience, bacteria important in foods have been arbitrarily divided into several groups on the basis of similarities in certain characteristics. This grouping does not have any taxonomic significance. Some of these groups and their importance in foods are listed here:

1. **Lactic acid bacteria**

   Those bacteria that produce relatively large quantities of lactic acid from carbohydrates. Include species mainly from genera Lactococcus, Leuconostoc, Pediococcus, Lactobacillus and Streptococcus thermophilus.

2. **Acetic acid bacteria**

   Those bacteria that produce acetic acid, such as Acetobacter aceti.

3. **Propionic acid bacteria**

   Those bacteria that produce propionic acid and are used in dairy fermentation. Include species such as Propionibacterium freudenreichii.

4. **Butyric acid bacteria**

   Those bacteria that produce butyric acid in relatively large amounts. Some Clostridium spp., such as Clostridium butyricum.

5. **Proteolytic bacteria**

   Those bacteria that are capable of hydrolyzing proteins due to production of extracellular proteinases. Species in genera Micrococcus, Staphylocccus, Bacillus, Clostridium, Pseudomonas, Alteromonas, Flavobacerium, and Alcaligenes; some in Enterobacteriaceae and Brevibacterium are also included in this group.

6. **Lipolytic bacteria**

   Able to hydrolyze triglycerides due to production of extracellular lipases. Species in genera Micrococcus, Staphylococcus, Serration, Pseudomonas, Alteromonas, Alcaligenes and Flavobacterium are included in this group.

7. **Saccharolytic bacteria**

   Able to hydrolyze complex carbohydrates. Include some species in genera Bacillus, Clostridium, Aeromonas, Pseudomonas, and Enterobacter.

8. **Thermophillic bacteria**

   Able to grow at 500 C and above. Include some species from genera Bacillus, Clostridium, Pediococcus, Streptococcus, and Lactobacillus.

9. **Psychrotrophic bacteria**

   Able to grow at refrigerated temperature (<50 C). Include some species of Pseudomonas, Alteromonas, Alcaligenes, Flavobacterium, Serratia, Bacillus, Clostridium, Lactobacillus, Leuconostoc, Listeria, Yersinia and Aeromonas.

10. **Thermoduric bacteria**

    Able to survive pasteurization temperature. Include some species of Micrococcus, Enterococcus, Lactobacillus, Pediococcus, Bacillus (spores) and Clostridium (spores).

11. **Halotolerant bacteria**

    Able to survive high salt concentrations (>10%). Include some species of Bacillus, Micrococcus, Staphylococcus, Pediococcus, Vibrio Streptococcus, Clostridium and Corynebacterium.

12. **Aciduric bacteria**

    Able to survive at low pH (below 4.0). Include some species of Lactobacillus, Pediococcus, Lactococcus, Enterococcus and Streptococcus.

13. **Osmophilic bacteria**

    Can grow at a relatively higher osmotic pressure (environment) than other bacteria. Some species from genera Staphylococcus, Leuconostoc, and Lactobacillus are included in this group. They are much less osmophilic than yeasts and molds.

14. **Gas-producing bacteria**

    Produce gas (CO2, H2, and H2S) during metabolism of nutrients. Include spices from genera Leuconostoc, Lactobacillus, Brevibacterium and Escherichia.

15. **Slime producers**

    Produce slime due to synthesis of polysaccharides. Include some species or strains of Xanthomonas, Leuconostoc, Alcaligenes, Enterobacter, Lactococcus, and Lactobacillus.

16. **Spore formers**

    Ability to produce spore. Include Bacillus, Clostridium and Desulfotomaculum spp. They are again divided into aerobic, anaerobic, flat sour thermophilic and sulfide-producing spore formers.

17. **Aerobes**

    Require oxygen for growth and multiplication. Species of Pseudomonas, Bacillus, and Flavobacterium are included in this group.

18. **Anaerobes**

    Cannot grow in the presence of oxygen. Include species of Clostridium.

19. **Facultative anaerobes**

    Able to grow both in the presence and in absence of oxygen. Lactobacillus, Pediococcus, Leuconostoc, enteric pathogens, some species of Bacillus, Serratia, and coliforms are included in this group.

20. **Coliforms**

    Include mainly species from Escherichia, Enterobacter, Citrobacter, and Klebsiella, and used as index of sanitation.

21. **Fecal coliforms**

    Include mainly Escherichia coli. Also used as index of sanitation.

22. **Enteric pathogens**

    Include pathogenic Salmonella, Shigella, Campylobacter, Yersinia, Escherichia, Vibrio, Listeria, Hepatitis A, and others that can cause gastrointestinal infection.

Fungi (Yeast and Molds)

Yeast

Yeasts are important in food due to their ability to cause spoilage. Many are also used in food bioprocessing. Some are used to produce food additives. Several important genera are briefly described below:

1. **Saccharomyces**: Cells are round, oval, or elongated. It is the most important genus and contains heterogeneous groups. Saccharomyces cerevisiae variants are used in baking for leavening of bread and in alcoholic fermentation. They are also involved in spoilage of food with the production of alcohol and CO_2.
2. **Pichia**: They are oval to cylindrical cells and form pellicle in beer, wine, and brine to cause spoilage. Some are also used in oriental food fermentation. Species: Pichia membranaefaciens.
3. **Rhodotorula**: They are pigment (red, pink or yellow) forming yeasts and can cause discoloration of foods, such as in meat, fish, and sauerkraut. Species Rhodotorula glutinis.

4. **Torulopsis**: They have spherical to oval structure. They cause spoilage of milk due to the ability to ferment lactose (Torulopsis sphaerica). They also spoil fruit juice concentrates and acid foods.
5. **Candida**: Many spoil foods with high acid, salt, and sugar and form pellicle on the surface of liquids. Some can cause rancidity in butter and dairy products (Candida lipolytica).
6. **Zygosaccharomyces**: Involved in spoilage of foods, containing high sugar/ salt levels ex. honey, sirups, molasses, soy sauce. (Zygosaccharomyces nussbaumeri). These yeasts are termed osmophilic, because they can grow in high concentrations of solutes.

Molds

- Molds are important in food because they can grow in conditions in which many bacteria cannot, such as low pH, low water activity (a w), and high osmotic pressure.
- They are important spoilage microorganisms.
- Many strains also produce mycotoxins and have been implicated in food borne intoxication.
- Many are used in food bioprocessing.
- Finally, many are used to produce food additives and enzymes.

Some of the most common genera of molds found in food are listed here:

1. **Aspergillus**

   They are widely distributed and contain many species that are important in food. They have septate hyphae and produce a sexual spores (black color) or conidia. Many are xerophilic (able to grow in low Aw) and can grow in grains, causing spoilage. They are also involved in spoilage of foods such as jams, cured ham, nuts, and fruits and vegetables (rot). Some species/strains produce mycotoxin (e.g., Aspergillus flavus produces aflatoxin). Many species/strains are also used in food and food additive processing. Aspergillus oryzae is used to hydrolyze starch by alpha-amylase in the production of sake. Aspergillus niger is used to process citric acid from sucrose and to produce enzymes like-galactosidase.

2. **Alternaria**

   They are also septate and form dark-brown colored many celled conidia on the conidiophere. They cause rot in tomatoes and rancid flavor in dairy products. Species: Alternaria tenuis.

3. **Geotrichum**

   The hyphae are septate and form rectangular asexual arthrospores (oidia). They grow forming a yeast like, cottony, creamy colony. They establish easily in equipment and often grow on dairy products (also known as dairy mold). Species: Geotrichum candidum.

4. **Mucor**

   They are widely distributed. They have nonseptate hyphae and produce sporangiophores. They produce cottony colonies. Some species are used in food fermentation and production of enzymes. They cause spoilage of vegetables. Species: Mucor rouxii.

5. **Penicillium**

   They are widely distributed and contain many species. They have septate hyphae and form conidiophore on a blue-green, brush like conidia head. Some species are used in food production, such as 17 Introduction to Food Microbiology Penicillium roquefortii and Penicillium camembertii in cheese. Many species cause fungal rot in fruits and vegetables.

6. **Rhizopus**

   The hyphae are aseptate and form sporangiophores in sporangium. They are involved in the spoilage of many fruits and vegetables. Rhizopus stolonifer is the common black bread mold.

Algae

- Algae are a very diverse group of generally simple unicellular or multicellular eukaryotic organisms. Most of them are autotrophic which means that they can harvest carbon dioxide from the atmosphere and convert it to organic matter.
- They inherited their photosynthetic apparatus from cyanobacteria.
- Cyanobacteria are sometimes called blue-green algae but they are prokaryotic organisms and are not true algae.
- Some cyanobacterial species are used as food as well.
- Seaweeds are edible algae that have been used for centuries as food in many coastal regions all over the world. They may belong to one of three groups of multicellular algae: red, green or brown. They are usually of marine origin since freshwater algae are often poisonous.
- Algae are of excellent nutritional value since they contain complete protein (in contrast to plant food harvested on land), fiber, and sometimes-high levels of omega-3 fatty acids.
- Algae are also rich in many vitamins, such as A, C, B_1, B_2, B_3 and B_6, as well as minerals, such as iodine, calcium, potassium, magnesium and iron. They are consumed both cooked, dried and raw.
- Cultivated microalgae and cyanobacteria such as *Spirulina* and *Chlorella* are sold as nutritional supplements.
- Hydrocolloids such as agar, alginate and carrageenan are isolated from wild and cultivated algae and used as additives in the food industry for their emulsifying and thickening properties.

Parasites

- A number of parasitic worms can also be transmitted by food to cause diseases in humans.

- Cestodes are flatworms that inhabit the intestinal tract, heart, and lungs of animals. Beef, swine, dogs and other canine species, bears, and fish can all harbor tapeworms and flatworms, which can be transmitted to and can infect humans.
- Trematodes are non-segmented flatworms that possess a mouth and oral sucker and depend on a snail as an intermediate host before infecting humans by being ingested in drinking water or aquatic plants. Intestinal flukes, pyriform worms from fish, sheep and Chinese liver flukes, and oriental lung flukes are all examples of food-transmitted parasites.
- Nematodes or true roundworms also can be transmitted from animals to humans. Eggs carried in excrement from roaches and dung beetles ingested by cattle, sheep and hogs contaminate humans. Trichinosis is an inflammation of the muscle tissue caused by ingesting the worm Trichinella spiralis. Pork is the most common vector. Capillary worms, whipworms, and pinworms are other examples of nematode parasites.
- Protozoa are microscopic single-celled animals, which can be taken in with food or water to cause human illness. Entamoeba histolytica, Toxoplasma gondii, Balantidium coli, and Giardia lamblia are the most common food borne protozoan parasites.

3. Factors Affecting the Growth of Microbes

Factors affecting microbial growth in food are:

1. **Intrinsic factors:** These are inherent in the food. They include: pH, water activity, oxidation reduction potential, nutrient content, antimicrobial contents, biological structure
2. **Extrinsic factors:** Are factors external to the food that affect microbial growth.

1. Intrinsic Factors

a) **Hydrogen ion concentration (pH):**

- Most bacteria grow best at neutral or weakly alkaline pH usually between 6.8 and 7.5.
- Some bacteria can grow within a narrow pH range of 4.5 and 9.0, e.g. Salmonella.
- Other microorganisms especially yeasts, molds, and some bacteria grow within a wide pH range, e.g. molds grow between 1.5 to 11.0, while yeasts grow between 1.5 and 8.5.

b) **pH values of some food products:**

- Beef 5.1 - 6.2
- Chicken 6.2 – 6.4
- Milk 6.3 – 6.8
- Cheese 4.9 - 5.9 Fish 6.6 - 6.8
- Oyster 4.8 - 6.3

- Fruits < 4.5 (most < 3.5)
- Vegetables 3.0 – 6.1

c) **Moisture content:**
- The amount of free water is important for growth of microorganisms.
- If there is, lack of free water microorganisms will not grow.
- Water activity is defined as the vapor pressure of a food substance to that of water at the same temperature. (Aw = VPFood/VPWater)
- Growth of microorganisms is greatly affected by the level of water activity (Aw) in the food.
- Inhibition of growth occurs if the water activity for food is lowered beyond an organism's minimum level of water activity that is necessary for growth.
- Microorganisms have varying minimum water activity requirements that supports their growth in food.

d) **Nutrients content of the food:**
- Microorganisms require proteins, carbohydrates, lipids, water, energy, nitrogen, Sulphur, phosphorus, vitamins, and minerals for growth.
- Various foods have specific nutrients that help in microbial growth.
- Foods such as milk, meat and eggs contain a number of nutrients that are required by microorganisms.
- These foods are hence susceptible to microbial spoilage.

e) **Antimicrobial substances:**
- Antimicrobial substances in food inhibit microbial growth.
- Various foods have inherent antimicrobial substances that prevent (inhibit) microbial attack.
- Such inhibitors are like lactenin and anti- coliform factors in milk and lysozyme in eggs.

f) **Biological structures:**
- Some foods have biological structures that prevent microbial entry.
- For example, meat has fascia, skin and other membranes that prevent microbial entry.
- Eggs have shell and inner membranes that prevent yolk and egg white from infection.

2. Extrinsic Factors

a) **Temperature:**
- The growth of microorganisms is affected by the environmental temperatures.
- Various microorganisms are able to grow at certain temperatures and not others.

- Bacteria can therefore be divided into the following groups depending upon their optimum temperature of growth.
  i) **Psychrophilic microorganisms:** These grow best at about 20°C but also down to -10°C in unfrozen media. Psychrophilic bacteria can cause food spoilage at low temperatures. Several of the microorganisms found in the soil and water belong to this group.
  ii) **Mesophilic bacteria:** These organisms grow between 25°C and 40°C, with an optimum growth temperature close to 37°C. Some such as Pseudomonas aeruginosa may grow at even lower temperatures between 5- 43°C. None of the mesophilic bacteria are able to grow below 5°C or above 45°C. Most pathogenic bacteria belong to this group.
  iii) **Thermophilic bacteria:** These grow at temperatures above 45°C. Often their optimum growth temperatures are between 50°C and 70°C. Growth of some bacteria occur at 80°C. Bacteria in this group are mainly spore formers and are of importance in the food industry especially in processed foods.

  The effect of temperature on microbial growth also depends upon other environmental conditions such as Growth factors in the nutrient medium, pH of the food, and Water activity.

b) Concentration of gases in the environment:

- This relates to the presence and concentration of gases in the food environment.
- Various microorganisms require for either growth, high oxygen tension (aerobic), low oxygen tension (micro aerobic) or absence of oxygen (anaerobic).
- Some microorganisms may grow either in high oxygen tension, or in the absence of oxygen (facultative anaerobes).
- Anaerobic or facultatively anaerobic spore formers are most likely to grow in canned foods.
- Microaerophilic bacteria are most likely to grow in vacuum-packed foods since they have low oxygen tension.
- Aerobic bacteria are likely to grow on the surface of raw meat.
- Aerobic molds will grow in insufficiently dried or salted products.

c) **Relative humidity:**

- Relative humidity is the amount of moisture in the atmosphere or food environment.
- Foods with low water activity placed at high humidity environment take up water, increase their water activity and get spoiled easily.
- For example, dry grains stored in an environment with high humidity will take up water and undergo mold spoilage.

4. Beneficial Role of Microorganisms

- The microorganisms occurring on and in foods are divided into three groups: bacteria yeast and molds.
- Molds are generally concerned in the food spoilage; their use in the food industry is limited.
- Yeasts are the most widely used microbes in the food industry due to their ability to ferment sugars to ethanol and carbon dioxide. Some types of yeast, like baker's yeasts are grown industrially, and some may be used as protein sources, mainly in animal feed.
- Bacteria are important in food microbiology. They may be divided into groups according to the fermentation product, e.g. lactic acid bacteria, propionic acid bacteria, acetic acid bacteria.
- Bearing in mind the food constituent attacked, prolytic, saccharolytic and lipolytic bacteria may be distinguished.
- Microorganisms are essential for the production of foods such as cheese, yogurt, bread, beer, wine and other fermented foods.
- Fermentation is one of the methods to preserve food and alter its quality.
- Yeast, especially Saccharomyces cerevisiae, is used to leaven bread, brew beer and make wine.
- Certain bacteria, including lactic acid bacteria, are used to make yogurt, cheese, hot sauce, pickles, fermented sausages and dishes.
- A common effect of these fermentations is that the food product is less hospitable to other microorganisms, including pathogens and spoilage-causing microorganisms, thus extending the food's shelf life.
- Some cheese varieties also require molds to ripen and develop their characteristic flavors.

FOOD CONTAMINATION AND SPOILAGE

Introduction

All food should be safe and free from contamination and spoilage at all points in its journey from its source until it reaches the consumers. However, food contamination is a serious public health problem in Ethiopia, resulting in food borne diseases that affect many people every year. Hence, awareness of potential sources of food contamination is an important component of good nutrition and good health. In this study session we are going to concentrate on food contamination by microorganisms, chemicals and physical factors.

Food may be contaminated by different microorganisms or by chemicals that can cause health problems for anyone who eats it.

Routes of Microbial Contamination of Food

Bacteria are a major source of microbial contamination of food, i.e. the undesired presence in food of harmful microorganisms or the harmful substances they produce. Viruses, parasites and fungi are also able to contaminate food and cause foodborne illnesses in humans. Microorganisms can enter food through different routes.

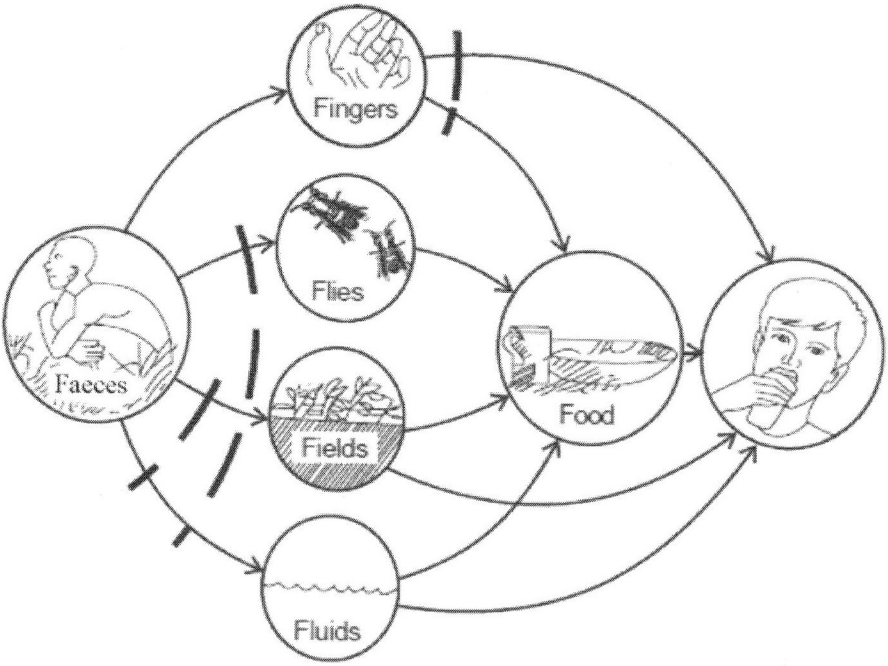

Look at Figure and notice that eating contaminated food is a component of many transmission routes. The most common routes of entry are discussed below.

Air and Dust

Microorganisms are found everywhere in our environment. Many types can be found in air and dust, and can contaminate food at any time during food preparation or when food is left uncovered (see below Figure). Imagine a kitchen where food is prepared and stored in rural communities, and think how easily microorganisms in the air and dust could contaminate the food.

Fig. Butter for sale in uncovered containers open to the air.

Soil, Water and Plants

Many microorganisms present in soil and water may contaminate foods. Microorganisms also grow on plants and can contaminate food if care is not taken to remove them by washing or inactivate them by cooking. Soil is a particularly rich source of *Clostridium* bacteria. Water may be contaminated by faeces. Plants may also be contaminated by faeces if untreated sewage has been used as a fertilizer.

Gastrointestinal Tract

The intestines of all humans and animals are full of microorganisms, some of which are beneficial but others are pathogenic. Bacterial pathogens such as Salmonella, Campylobacter and Escherichia coli (strain O157:H7) are common examples. Contamination of foods by faecal material is the major cause of food poisoning events. This includes indirect contamination, for example from people's hands if they prepare food without washing their hands after visiting the latrine/toilet.

Escherichia coli (abbreviated to E.coli) exists in many harmless varieties or 'strains', but some strains are pathogenic. The strain called E.coli O157:H7 causes a potentially fatal foodborne disease in humans.

Animals

Many foodborne microorganisms are present in healthy animals raised for food, usually in their intestines, hides, feathers, etc. Meat and poultry carcasses can be contaminated during slaughter by contact with small amounts of intestinal contents. For example, in animals slaughtered in rural communities without any safety measures, microorganisms present in the animals' intestines can easily contaminate the meat.

Animal hides are an important source of contamination of the general environment, the hands of meat workers, and skinned meat carcasses. Hides are a primary source of E.coli O157:H7 and Salmonella species, both of which cause sickness and diarrhoea. Hides become contaminated either because the outside of the hide is dirty, or because once removed from the animal, the inside of the hide is a good breeding place for microorganisms.

Animal Feeds

Animal feeds are a source of microorganisms, especially Salmonella, which can contaminate poultry and other farm animals. The organisms in dry animal feed spread throughout the local environment and may get on to animal hides, hair and feathers, as well as on people who handle the feeds.

Food Handlers

The term **food handler** can be applied to anyone who touches or handles food, and this includes people who process, transport, prepare, cook and serve food. The presence of microorganisms on the hands and outer garments of food handlers reflects the standard of hygiene in the environment and the individuals' personal hygiene (as you learned in earlier study sessions). The microorganisms transmitted to foods by food handlers may come from the hides of animals, soil, water, dust, gastrointestinal tracts and other environmental sources. In food preparation at home, foodborne microorganisms can be introduced from the unwashed hands of people who are infected by bacteria and viruses, and who cook and serve the food to family members.

Food Utensils

Food utensils are cutting boards, knives, spoons, bowls and other equipment used in food preparation, which may become contaminated during food processing and preparation. For example, in families where there is no access to running water, the food utensils may not be properly cleaned, stored and handled, and may become a major route of food contamination.

Cross-Contamination

Cross-contamination of food is the transfer of harmful microorganisms between food items and food contact surfaces. Prepared food, utensils and surfaces may become contaminated by raw food products and microorganisms. These can be transferred from one food to another by using the same knife, cutting board or other utensil without washing it between uses. A food that is fully cooked can become re-contaminated if it touches raw foods or contaminated surfaces or utensils that contain pathogens. For example, you should never:

- allow raw meat to touch cooked meat
- put cooked meat on a cutting board that has just been used for raw meat without cleaning it first (Figure below)
- store raw meat on a shelf above cooked meat so that it could leak blood and raw juices on to the cooked meat below.

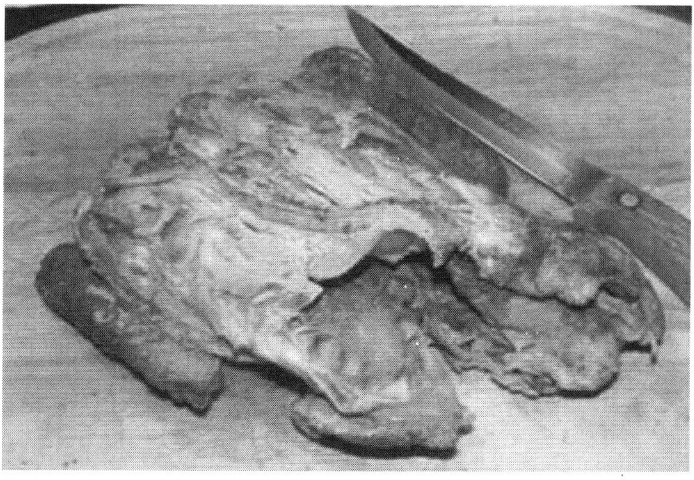

Never put cooked meat on an unclean cutting board that has been used to prepare raw meat.

Unsafe Temperature (Danger Zone Temprature)

An unsafe temperature for food storage is a major factor in food contamination. Many microorganisms need to multiply to a very large number before enough are present in food to cause disease in someone who eats it. However, if bacteria can have warm, moist conditions and an ample supply of nutrients, one bacterium can reproduce by dividing (on average) every half an hour and can produce 17 million bacteria in 12 hours! So, if you leave lightly contaminated food out overnight, it will be highly contaminated and infectious by the next day.

The Danger zone is the temperature range in which food-borne bacteria can grow. Food safety agencies such as The United States "Food Safety and inspection Service" (FSILS), define the danger zone as roughly 4*c to 60*c. The FSIS stipulates that potentially hazardous food should not be stored at temperatures in this range in order to prevent foodborne illness (for example, our refrigerators temperature must be kept below 4*c, and that food that remains in this zone for more than two hours should not be consumed. Foodborne micro organism grow much faster in the middle of the zone, at temperature between 21*c and 41*c

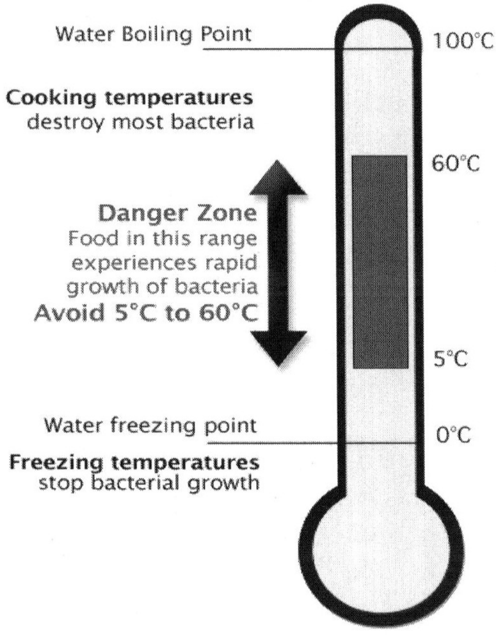

Poor Personal Hygiene

Poor personal hygiene of food handlers is another major factor in food contamination. The most important contaminants of food are the microorganisms excreted with faeces from the intestinal tract of humans. These pathogens are transferred to the food from faecal matter present on the hands.

We have already mentioned failure to wash hands after visiting a toilet as a source of food contamination. Can you suggest other times when food handlers should wash their hands?

Bad personal habits like scratching your hair and nose with your fingers also contributes to food contamination. Sneezing and coughing spreads contaminants and microorganisms through the air and onto uncovered food, and onto surfaces and hands that can transfer the infectious agents into food.

Pests

Foods can be damaged and also contaminated by pests. Many stored grains are lost through the damage done by pests, including termites (*mist*), beetles, locusts, cockroaches, flies and rodents such as rats and mice. Pests can damage and contaminate foods in various ways, such as boring into and feeding on the insides of grains, or tunnelling into stems and roots of food plants. For example, weevils cause large losses of stored grains, especially in warm and humid conditions such as in lowland areas of Ethiopia.

Pests also damage the protective skin of foods allowing microorganisms to get inside the food and causing it to rot more quickly. Pests can pollute food with their excreta, and with bodies and body fragments when they die. They also transfer microorganisms on to food while walking on it (Figure below). Flies and cockroaches readily move between wastes and foods, transporting microorganisms with them as they go.

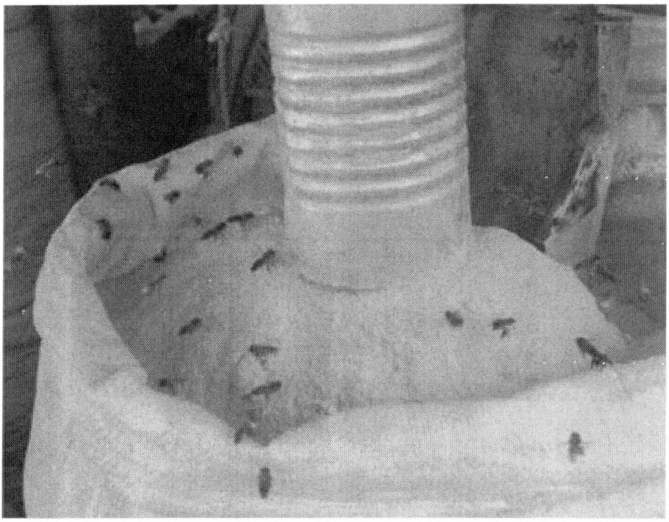

Insects can leave dirt, excreta and possibly pathogenic microorganisms if they are allowed to crawl on food.

Avoiding Food Contamination

You now know that food can be contaminated from sources in the natural environment, people, food preparation surfaces and utensils, raw and uncooked food, animals, pests, and waste material. To prevent contamination, food production and preparation operations need to be carefully controlled.

Microbial Food Contamination

Prevention of microbiological (often abbreviated to 'microbial') contamination is an important function in food preparation.

Avoiding Microbial Food Contamination

Food handlers should follow these strategies:

- Thorough hand washing before and during food preparation, especially after using the toilet, and handling raw food or waste.
- Soap/ash sanitizer and clean water should be available for hand washing at convenient locations.
- Sick food handlers should not prepare food! One sick person can cause a foodborne disease outbreak, particularly where people are in crowded or unsanitary living conditions.
- Raw and cooked foods should be separated, because raw foods are a source of microorganisms and can re contaminate prepared foods.

Chemical Contamination Of Food

Attention also needs to be given to possible chemical contamination of food. Food can be contaminated through the misuse or mistaken handling of chemicals, including pesticides, bleach and other cleaning materials. All chemicals (detergent, disinfectant, sanitiser) used in the food preparation area should be removed before food preparation begins, to prevent any chemical contamination of the food.

Other possible sources of chemical contamination are:

- reusing containers which have been used for chemicals (Figure below)
- using chemical sprays (e.g. to kill cockroaches) in areas where food is exposed
- accidentally adding chemicals which have a texture similar to table salt or sugar during food preparation; they should always be stored separately.

The can that is being used to scoop salt from this sack has previously been used for insecticide.

Physical Contamination of Food

Physical contaminants include stones, pieces of glass, and metal. Physical contamination can occur at any stage of the food chain: for example, stones, bones, twigs, pieces of shell or foreign objects can enter food during handling and preparation. These materials should be removed, if possible, for example by sieving or picking out the items with clean fingers.

Food Spoilage

Food spoilage is the process of change in the physical and chemical properties of the food so that it becomes unfit for consumption. Food spoilage is any undesirable change in food. Most natural foods have a limited life: for example, fish, meat, milk and bread are perishable foods, which means they have a short storage life and they easily spoil. Other foods also decompose eventually, even though they keep for a considerably longer time. The main cause of food spoilage is invasion by microorganisms such as fungi and bacteria.

Microbial Spoilage

Microbial spoilage is caused by microorganisms like fungi (moulds, yeasts) and bacteria. They spoil food by growing in it and producing substances that change the colour, texture and odour of the food. Eventually the food will be unfit for human consumption.

When food is covered with a furry growth and becomes soft and smells bad, the spoilage is caused by the growth of moulds and yeasts. Microbial spoilage by moulds and yeasts includes souring of milk, growth of mould on bread and rotting of fruit and vegetables. These organisms are rarely harmful to humans, but bacterial contamination is often more dangerous because the food does not always look bad, even if it is severely infected. When microorganisms get access to food, they utilise the nutrients found in it and their numbers rapidly increase. They change the food's flavour and synthesise new compounds that can be harmful to humans. Food spoilage directly affects the colour, taste, odour and consistency or texture of food, and it may become dangerous to eat. The presence of a bad odour or smell coming from food is an indication that it may be unsafe. But remember that not all unsafe food smells bad.

What is the difference between food contamination and food spoilage?

The term **contact spoilage** is used when microbial spoilage is the result of direct contact or touching between the food and any contaminated or unclean surface such as shelves, food preparation boards or unwashed hands. It also includes food-to-food contact, for example between cooked meat and raw meat or between rotting fruit and sound fruit.

Physical Spoilage

Physical spoilage is due to physical damage to food during harvesting, processing or distribution. The damage increases the chance of chemical or microbial spoilage and contamination because the protective outer layer of the food is bruised or broken and microorganisms can enter the foodstuff more easily. For example you may have noticed that when an apple skin is damaged, the apple rots more quickly.

Chemical Spoilage

Chemical reactions in food are responsible for changes in the colour and flavour of foods during processing and storage. Foods are of best quality when they are fresh, but after fruits and vegetables are harvested, or animals are slaughtered, chemical changes begin automatically within the foods and lead to deterioration in quality. Fats break down and become rancid (smell bad), and naturally-occurring enzymes promote major chemical changes in foods as they age.

Enzymic Spoilage (Autolysis)

Every living organism uses specialised proteins called **enzymes** to drive the chemical reactions in its cells. After death, enzymes play a role in the decomposition of once-living tissue, in a process called **autolysis** (self-destruction) or **enzymic spoilage**. For example, some enzymes in a tomato help it to ripen, but other enzymes cause it to decay (Figure below). Once enzymic spoilage is under way, it produces damage to the tomato skin, so moulds can begin to can attack it as well, speeding the process of decay.

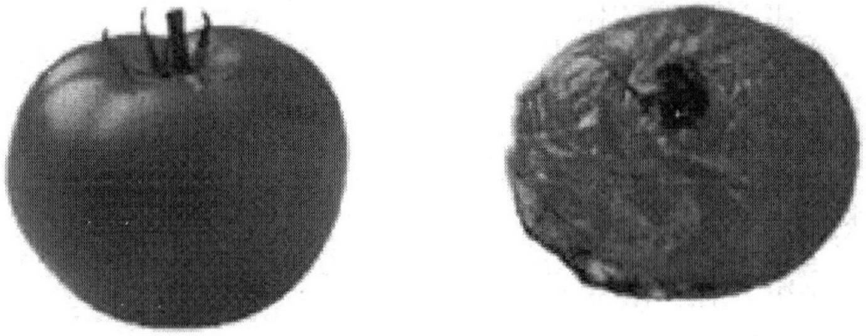

Role of enzymes in tomato spoilage: the tomato on the right has also been attacked by fungi (moulds), speeding its decay.

Enzymic Browning

When the cells of fruits and vegetables such as apples, potatoes, bananas and avocado are cut and exposed to the air, enzymes present in the cells bring about a chemical reaction in which colourless compounds are converted into brown-coloured compounds. This is called **enzymic browning.** If the food is cooked very soon after cutting, the enzymes are destroyed by heat and the browning does not occur. For example, apples are prone to discolouration if cut open when raw, but when cooked they do not go brown.

Appearance of Spoiled Food

Spoiled food is generally more a problem of appearance than a problem of disease causing. In food spoilage, the changes in appearance or texture of the food, such as rottenness, softness and change in colour, taste or odour are usually obvious, whereas in contaminated food such characteristics may not be noticed. A large majority of the microorganisms responsible for food spoilage are not pathogenic to humans. However, you should advise people in your community that they should not eat food that is spoiled because it is not nutritious and may make them sick (cause vomiting).

Factors Affecting Food Spoilage

Finally, we turn to the factors that can increase or delay the process of food spoilage. They include its water content, environmental conditions, packaging and storage.

Water Content

The amount of water available in a food can be described in terms of the water activity (a_w).

- Can you recall the a_W of pure water?
- The water activity of most fresh foods is 0.99. This means that they have a very high water content and can support a lot of microbial growth.
- Meat is traditionally dried by adding table salt to it. Can you suggest why salting enables the meat to be stored for a long time?

Environmental Conditions

No matter whether food is fresh or processed, the rate of its deterioration or spoilage is influenced by the environment to which it is exposed. The exposure of food to oxygen, light, warmth or even small amounts of moisture can often trigger a series of damaging chemical and/or microbial reactions. Changing the environment can help to delay spoilage. For example, storing foods at low temperatures reduces spoilage because both microbial and enzymic decay is faster at higher temperatures.

Packaging and Storage

Packaging is a means of safeguarding food when it is raw, or after it has been processed or prepared. It helps to protect food against harmful contaminants in the environment or conditions that promote food spoilage including light, oxygen and moisture. The type of packaging is a key factor in ensuring that the food is protected. Packaging of foods in cans, jars, cartons, plastics or paper also serves to ensure food safety if it is intact, because it provides protection against the entry of microorganisms, dust, dirt, insects, chemicals and foreign material.

SANITARY PROCEDURE DURING FOOD HANDLING

Clean Procedure Followed During Food Handling

The treatment of food can occur during:

- Cooking
- Cooling
- Hot holding
- Preparation
- Purchase
- Receipt
- Re-warming

- Serving
- Storage

General Safe Food Handling Tips

- Do not wipe your hands on your garments as this can without much of a stretch exchange microorganisms and microbes.
- Use paper towels to tidy up during food planning and serving.
- Change gloves, utensils and dishes when evolving capacities. For example utilize one sets of gloves for dealing with crude meat, and another pair taking care of new vegetables.
- Never run in food creation or administration zones
- Try to have only one individual serve food that is going to be eaten.
- Prepare precooked solidified nourishments precisely as the headings/ guidelines on the bundling state.
- Have nourishments prepared no longer than should be expected before spending time in jail.
- Prepare and cook just as much food as you expect to utilize.
- Wash and disinfect flatware or different utensils, which tumble to the floor.
- Do not taste nourishments with any utensil utilized either to blend or mix food.
- Pick up and hold all flatware by the handles.
- Store flatware away from dust.
- Be cautious while lifting tops from hot food.
- Turn handles of pans from the front of the oven when cooking.

Getting Ready to Eat Food

At whatever point conceivable consistently attempt to deal with any food things that are going to be eaten, with an utensil (for example utensils) as opposed to your uncovered hands.

Hand Washing

Clean hands are basic for working in a kitchen situation. It's simple for microbes to spread from the food we contact to entryway handles, plates, cutlery, etc. Hand washing is perhaps the most ideal approaches to forestall the spread of germs between individuals.

When washing your hands attempt to;

- Use a cleanser container instead of a bar of cleanser.
- Wash in a sink that has hot and cold running water.
- Wash in a sink that is independent from one that is utilized to wash staple and utensils.
- Dry your hands with paper towels.

Wash your hands after:

- Starting work
- Using the latrine
- Handling crude and cooked nourishments
- Taking breaks
- Eating
- Drinking
- Smoking
- Coughing, wheezing or cleaning out their nose
- Touching your hair
- Playing with pets or dealing with creatures
- Scratching
- Handling deny or squander materials
- Handling cleaning synthetic concoctions
- Method to washing your hands appropriately
- Wet your hands
- Rub your hands and wrists with cleanser
- Lather the cleanser for 20 seconds
- Rinse altogether
- Dry with paper towels or a hot air dryer (recall that wet hands can convey and move a larger number of germs than dry ones)
- Turn of the taps with your elbows (if conceivable) or utilize a paper towel to do as such.
- Dealing with DISHES, CROCKERY AND CUTLERY
- Try not to contact any piece of a dish or plate which will come into contact with an individual's food or mouth.
- Pick up cups and mugs by their handles, your fingers ought to be outside cups.
- Place teaspoons so they project from a dish.
- Pull out dispensable cups from the base of a cylinder, this keeps your fingers from going inside the cup.
- Do not use plates which have gotten split or chipped.

Food Handling and Storage

- Date all food when it is gotten. Ensure that more established nourishments are spent first to keep up sustenance and newness.
- Seal bundles after they are opened and refrigerate uneaten food to maintain a strategic distance from decay.
- Keep hot nourishments hot and cold food sources cold. Hot nourishments

ought to have an inside temperature of 600C/1400F. Refrigerate new nourishments (e.g., meat, fish, poultry, milk and milk choices, eggs and egg items) directly after they are purchased or gotten. Cold nourishments must be kept at 40C/400F or less.

- Place a thermometer in every fridge. Check temperatures every day to guarantee that nourishments are kept at an appropriate temperature of 40C/400F or less.
- Keep crude food away from cooked or prepared to eat nourishments.
- Keep harmful and noxious substances separate from food and far from kids and youth.
- Keep re-usable basic food item sacks clean and wash them consistently (whenever utilized).
- Always wash your hands before dealing with food.

Food Preparation

All program locales need a sink in the arrangement region for hand washing just, with hot and cold running water, cleanser and paper towels in distributors. If it's not too much trouble contact your neighborhood general wellbeing unit if your program site doesn't have a sink in the readiness region to guarantee you are taking care of food securely.

- Clean all work surfaces after each utilization.
- Wash food surfaces with hot, sudsy water, at that point wash.
- Sanitize food work surfaces by splashing a safe disinfecting arrangement. Contact your nearby general wellbeing unit or educational committee for disinfection rehearses.
- Wash kitchen towels, wipes, and materials frequently. Microscopic organisms can live and develop on these things.
- Wash cutting sheets and blades in hot foamy water. Wash and disinfect utensils subsequent to cutting up crude meat, poultry, and fish and before utilizing the utensils to plan other food.
- Keep crude meat, poultry and fish and their juices from different nourishments.
- Thaw food in cooler, microwave, or under virus running water. Try not to defrost food on kitchen counter.
- Prepare all nourishments as near spending time in jail as could reasonably be expected.
- Cook meat, fish, poultry and eggs, until they arrive at a safe interior temperature and are totally cooked. Store cold nourishments beneath 40C/400F, hot food sources above 600C/1400F if not serving them immediately.
- Chill prepared food rapidly so it invests minimal measure of energy conceivable in the "temperature peril zone". Appropriate capacity keeps

extras at their freshest, longer. Refrigerate or freeze all extras inside two hours in the wake of cooking to limit the opportunity of microscopic organisms developing.

Hot Holding and Cold Holding Food

In the event that you are holding nourishments for administration, for example, on a smorgasbord line or in a cafeteria, at that point attempt to keep hot nourishments hot and cold food sources cold. Hot holding gear alongside scraping dishes, slow cookers, and warming plate all assistance to keep prepared to eat food out of the peril zone. The entirety of this hardware is for hot holding just, and ought not be utilized to warm or prepare food.

Tips:

- Preheat hot holding hardware before you put any food in it. In the event that you don't, at that point you will place food into cold gear, which energize microscopic organisms' development.
- Limit the hot holding of food to a limit of two hours.
- To circulate the warmth uniformly, make a point to mix the food at ordinary stretches.
- Keep the food secured, this holds the warmth as well as prevents pollutes from falling into food.
- Bring out the food as close as conceivable to the hour of administration.
- Keep platters refrigerated until the time has come to warm them in the mood for serving.

SAFE FOOD HANDLER

Personal Hygiene

1. Health Status

- Food handlers shall undergo a medical examination by a registered medical practitioner annually to ensure that they are free from any infectious & other communicable diseases.
- A record of these examinations shall be maintained in the Performa as prescribed under Food Safety & Standards Act, 2006.
- Food handlers shall be inoculated against the enteric group of diseases as per the recommended schedule of the vaccine & records shall be maintained.
- Medical examination to be concluded includes – Physical examination, Eye test, Skin examination
- Any test required to confirm any communicable or infectious disease which the person suspected to be suffering from, on clinical examination.
- Vaccine to be inoculated against enteric group of disease, shall be decided by the medical practitioner according to the list as declared by the municipal corporation of that area.

2. Illness & Injury

- Food handlers suffering from of a disease shall not be allowed to handle food or material which comes in contact with food.
- Employees shall report the following conditions to the supervisor for possible exclusion from food handling areas - Jaundice, diarrhoea, vomiting, fever, sore throat with fever, visibly infected lesions, boils, cuts or sores & discharge from ears, eyes or nose.
- Medical examination of a food handler shall be carried out apart from the periodic medical examination, if clinically or epidemiologically indicated.
- Personnel with open cuts, wounds or burns shall be required to cover them with suitably water proof dressings before starting operation.
- Any lost dressing must be reported to supervisor immediately.
- The dressing should preferably be of bright colour & metal detectable.

3. Personal Cleanliness

- Food handlers shall maintain high degree of personal hygiene.
- Bathe or shower daily.
- They shall wear work clothing, head covering & footwear that is fit for the purpose, clean & in good condition (free from tears, rips or fraying material).
- Work wear shall provide adequate coverage to ensure that hair, beard, perspiration etc. cannot contaminate the product.
- Work wear should be free from buttons, with outside pockets above waist level. Zips or press-stud fastenings are acceptable.
- They should be laundered to standards and at intervals suitable for the intended use of the garments.
- Head cover should be worn first & footwear at the last, followed by sanitization.
- Protective clothing mandated for the food production areas shall not be used for any other purpose.
- Protective clothing includes – hair net, moustache net, glasses, ear plugs, gloves, aprons, foot wear.
- The aprons & dresses of food handlers kept in an ozonized cabinet or UV induced cabinets when handling sensitive products like pies, cakes.
- Where gloves are used for product contact, they shall be clean & in good condition.
- Working without gloves can be done provided there are necessary controls on periodic usage of disinfectants at work section & nature of the product being handled.
- Hair shall be kept neatly tied & fingernails shall be kept trimmed. Keep moustaches and beards trimmed and clean. Better yet, be clean-shaven.

- The food handlers shall prohibit the use of nail polish, false nails and false eyelashes; carrying of writing implements behind the ears.
- No strong perfume/lotion should be applied.
- Shoes worn outside food handling area shall not be allowed to enter food handling area.
- Shoes for use in processing areas shall be fully enclosed and made from non-absorbent materials.
- Street shoes either shall be changed or covered using foot cover.
- All people entering the food handling area shall wash their hands.
- Hand-washing notices should be posted in appropriate areas.
- Hands shall also be washed after –
    a) Handling non - food chemicals.
    b) Handling incompatible food products (such as raw versus cooked food) or contaminate material.
    c) Breaks, Eating, Drinking
    d) Coughing, sneezing, or blowing their nose.
    e) Using toilet facilities.
    f) Using cell phones
    g) Smoking
- As a good practice, cell phones should be used as minimum as possible (especially in high-risk areas) as they are also a source of contamination.
- Food handler should not handle soiled currency notes/cards to avoid contamination.
- Food handlers shall pass through air curtain to remove any lint or hair while leaving the changing room.

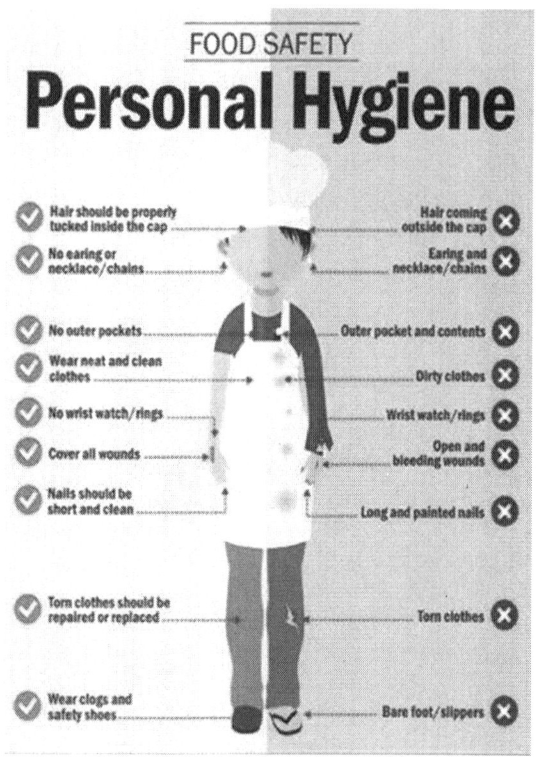

4. Personal Behaviour

- All food handlers shall follow a good personal behaviour.
- Any behaviour or unhygienic practice,

which could result in contamination of food, shall be prohibited in food handling areas. It includes –

a) Smoking
b) Chewing
c) Eating
d) Unprotected sneezing or coughing
e) Spitting

- Food handlers shall avoid certain hand habits such as scratching nose, running fingers through hair, rubbing eyes, ears & mouth, scratching beard or part of bodies.
- When unavoidable, hands shall be effectively washed before resuming work after such actions.
- Cover your mouth when you cough or sneeze, then wash your hands.
- Personal effect such as jewellery, watch, pins or other items should not be worn or brought into food handling areas if they pose threat to the safety and suitability of food. Avoid facial and/or body piercings; if you have them, do not touch them when you are at work.
- Food contact tool & equipment shall not be kept in personal lockers. Processing equipment (for example, refrigerators and freezers) should not be used for personal storage (such as storing lunches).
- Do not sit on worktables.

5. Visitors

1. Generally visitors should be discouraged from going inside the food handling areas.
2. Visitors when entering food manufacturing, cooking, preparation and storage or handling areas shall wear protective clothing, footwear.
3. Visitors shall adhere to the personal hygiene provisions as mandate for food handlers.

Visitor policy shall be documented

Visitor shall be given visitor card with restricted entry

Visitor shall wear protective clothing & footwear and shall adhere to food safety provisions as mandate.

Use of Gloves

- When used correctly, gloves can help protect foods against cross-contamination.
- When used incorrectly, they can spread contamination just as easily as bare hands.
- Health departments in some localities require the use of some kind of barrier between hands and foods that are ready to eat—that is, foods that will be served without further cooking.
- Gloves, tongs, and other serving implements, and bakery or deli tissue all can serve as barriers.
- To be sure to use gloves correctly, observe the following guidelines.

Guidelines: Using Disposable Gloves

1. Wash hands before putting on gloves or when changing to another pair. Gloves are not a substitute for proper handwashing.
2. Remove and discard gloves, wash hands, and change to a clean pair of gloves after handling one food item and before starting work on another. In particular, never to fail to change gloves after handling raw meat, poultry, or seafood. Gloves are for single use only. Remember that the purpose of wearing gloves is to avoid cross-contamination.
3. Change to a clean pair of gloves whenever gloves become torn, soiled, or contaminated by contact with an unsanitary surface.

Hand Washing

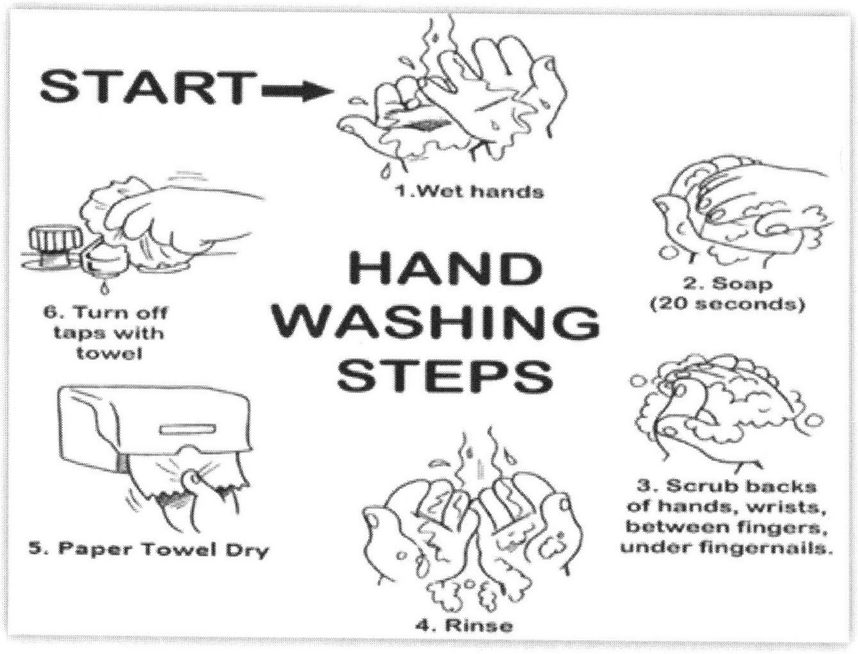

Procedure:

1. Wet your hands with hot running water. Cause the water as hot as you to can easily stand, yet at any rate 100°F (38°C).
2. Apply enough cleanser to make a decent foam.
3. Rub hands together altogether for 20 seconds or more, washing the hands as well as the wrists and the lower some portion of the lower arms.
4. Using a nailbrush, clean underneath the fingernails and between the fingers.
5. Rinse hands well under hot running water. On the off chance that conceivable, utilize a spotless paper towel to kill the water to abstain from sullying the hands by contact with dirtied spigots.
6. Dry hands with clean single-use paper towels or a warm-air hand dryer.

First Aid

- First aid is the first and immediate assistance given to any person suffering from either a minor or serious illness **or** injury, with care provided to preserve life, prevent the condition from worsening, or to promote recovery.
- It includes the complete treatment of minor conditions, such as applying a plaster to a cut.
- First aid is a combination of simple procedures and common sense.
- The aims of first aid are: To preserve life, To prevent further harm, Promote recovery

Cuts

- Cuts are open wounds through the skin.
- Normally the skin is under slight, constant tension as it covers the body.
- A cut is a forceful injury to the skin. Many people accidentally cut themselves with household or work items, yard tools, or when operating machinery.
- Most cuts are minor and home treatment is usually all that is needed.

Cuts can be caused by:

a) Blunt objects that tear or crush the skin (lacerations). These cuts are more common over bony areas, such as a finger, hand, knee, or foot, but they can occur anywhere on the body. Blunt object injuries usually cause more swelling and tissue damage and leave jagged edges, so problems with healing may occur.

b) Sharp-edged pointed objects pressing into and slicing the skin tissue (incised wounds). Sharp object injuries are more likely to cut deeper and damage underlying tissue.

c) Sharp-edged objects piercing the skin tissue (stab or puncture wounds).

A combination of blunt and sharp objects that tear, crush, and slice the skin tissue.

Types of Cuts

Some types of cuts are more serious and need medical evaluation and treatment. These more serious cuts include:

- Long or deep cuts.
- Cuts that open with movement of the body area, such as a cut over a joint. A cut over a joint may take a long time to heal because of the movement of the wound edges.
- Cuts that may scar and affect the appearance or function of a body area. A cut on an eyelid or lip, which does not heal well, may interfere with function or leave a noticeable scar.
- Cuts that remove all of the layers of the skin (avulsion injuries), such as slicing off the tip of a finger. An avulsion injury may take a long time to heal.
- Cuts that have damage to underlying tissues. Injuries to nerves, tendons, or joints are more common with cuts on the hands or feet. Slight swelling, bruising, and tenderness around a cut, bite, scrape, or puncture wound is normal. Swelling or bruising that begins within 30 minutes of the injury often means there is a large amount of bleeding or that damage to deeper tissues is present.
- Cuts over a possible broken bone. Bacteria can get into a cut over a broken bone and infect the bone.
- Cuts caused by a crushing injury. With this type of injury, the cut may have occurred when the skin split open from the force of the injury. The force of the injury may also damage underlying tissues and blood vessels. Crush injuries have a high risk of infection.
- Cuts with a known or suspected object, such as glass or wood, in the wound.
- Injury to the skin may also break small blood vessels under the skin and cause more swelling and bruising than you would expect.

When you have a cut:

- Stop the bleeding with direct pressure to the wound.
- Determine if other tissues, such as blood vessels, nerves, tendons, ligaments, joints, bones, or internal organs, have been injured.
- Determine if treatment by a doctor, such as stitches, staples, or skin adhesives, is needed.

- Clean the wound and remove any dirt or debris to prevent infections, both bacterial skin infections and tetanus ("lockjaw")
- Determine whether you need a tetanus shot.
- Cuts to the head or face may appear worse than they are and bleed a lot because of the good blood supply to this area. Controlling the bleeding will allow you to determine the seriousness of the injury.

Wounds

- An internal or external injury to living tissue caused by a cut, blow, or other impact, typically one in which the skin is cut or broken.
- In pathology, it specifically refers to a sharp injury, which damages the epidermis of the skin.
- Wounds are of two kinds; **open wounds and closed wounds.**
- The most common accidents resulting in open wounds are falls, mishandling of sharp objects, accidents with tools or machinery, and car accidents. An open wound (as in a knife cut) is a break in the skin or mucous membrane.
- A closed wound (a contusion or internal bleeding) is a bruise that damages the underlying tissue without breaking the skin (as in a black eye).

Open Wounds: Types

1. **Abrasions**

   An abrasion is a skin wound caused by rubbing or scraping the skin against a hard, rough surface.

   Bleeding in this type of wound is usually limited, but it is important that the skin be cleaned in order to guard against infection.

   Abrasions are often caused by a sliding fall onto a rough surface such as asphalt, tree bark or concrete.

2. **Incisions**

   An incision is a cut caused by a knife, the rough edge of metal, broken glass, a razor blade or some other sharp object.

   This type of wound generally bleeds rapidly and heavily. If the cut is deep, muscles, tendons and nerves may be damaged.

3. **Lacerations**

   A laceration is a jagged, irregular or blunt breaking or tearing of soft tissues, often resulting from mishandling tools and machinery and other accidents.

   Lacerations and incisions may appear linear (regular) or stellate (irregular).

   Bleeding from a laceration may be rapid and extensive.

   The term laceration is commonly misused in reference to incisions.

4. **Punctures**

   A puncture is a piercing wound that causes a small hole in the tissues.

   Such objects as splinter, nails, needles, ice picks and other pointed objects can produce puncture wounds.

   Even if external bleeding is slight, there may be serious internal bleeding resulting from internal damage to an organ (as in a gunshot wound).

   All puncture wounds require the attention of a health professional because of the danger of tetanus.

5. **Avulsions**

   An avulsion is a forcible tearing or partial tearing away of tissues.

   It occurs in such accidents as gunshot wounds, explosions, animal bites or other body-crushing injuries. Bleeding is heavy and rapid.

   If a body part (a finger, tooth or toe, for example) has been torn away in an accident, it should always be sent along with the victim to the hospital (placed on moistened gauze and cooled), since there is a possibility that it can be reattached.

6. Penetration Wounds

- An item, for example, a blade entering and coming out from the skin causes it.

Shut Wounds

1. Crush Injury

   It is brought about by an extraordinary or outrageous measure of power applied over an extensive stretch.

   A squash injury will be injury by an article that causes pressure of the body Slashes

   A profound cut or tear in skin or tissue.

   Lacerations might be profound or shallow, long or short, and wide or thin.

   Most cuts are the aftereffect of the skin hitting an item, or an article hitting the skin with power.

Cut Repair

- It is the demonstration of cleaning, planning, and shutting the injury.
- Minor cuts (shallow, little, not dying, and clean) may not require clinical consideration.
- Antibiotic treatment and a swathe might be all that is required. Be that as it may, most cuts do require fix.
- Cleaning and setting up a slash for fix is vital for forestalling disease and diminishing the presence of frightening.
- Cleaning washes away soil, yet additionally evacuates the germs that could trigger disease.

- Cleaning is done in a similar way paying little mind to the procedure that will be utilized for wound conclusion.
- Preparation is never really out rough edges so that scarring might be less recognizable.
- Preparation is done varying.

When is Lacerations Repair Needed?

The following are the recorded situations when you need cut fix

- Exposed muscle, fat, ligament or bone.
- Dirt and flotsam and jetsam in the injury, considerably in the wake of cleaning
- Feeling as though something is in the injury
- Bleeding proceeds subsequent to applying direct weight for 10-15 minutes
- Depth multiple/8 or 1/4 inch down
- Location on territory of high pressure
- Possible extraordinary scarring

HAZARD ANALYSIS CRITICAL CONTROL POINT (HACCP)

1. Introduction to HACCP

Implementing Hazard Analysis and Critical Control Point (HACCP) is crucial for any food manufacturing process. A HACCP plan covers the total supply chain, from inbound logistics, through storage, processing, sanitation and maintenance to the final use by the consumer. Across the operations, it must be ensured that procedures are available for internal logistics, processing specifications, working instructions, hygiene procedures and preventive maintenance plans. These procedures must cover start-ups, shutdown and unexpected stoppages during processing.

Hazard Analysis Critical Control Point (HACCP) is essential to carry out to identify the weakness of the production line and to suggest critical limits in compliance with legislation and therefore the preventive and corrective measures.

Though HACCP system was designed to aim zero defect products, yet it is not feasible to achieve 100% defect free products. However, it sets a goal to minimize the associated risks during production and subsequently reduce unacceptable unsafe products.

During implementation of HACCP, it is imperative to set controls at each point of the production line at which safety problems (physical, chemical and microbiological) are likely to occur.

A HACCP plan is required to be in place before initiating the HACCP system. A HACCP plan consists of 5 initial steps and 7 major HACCP principles.

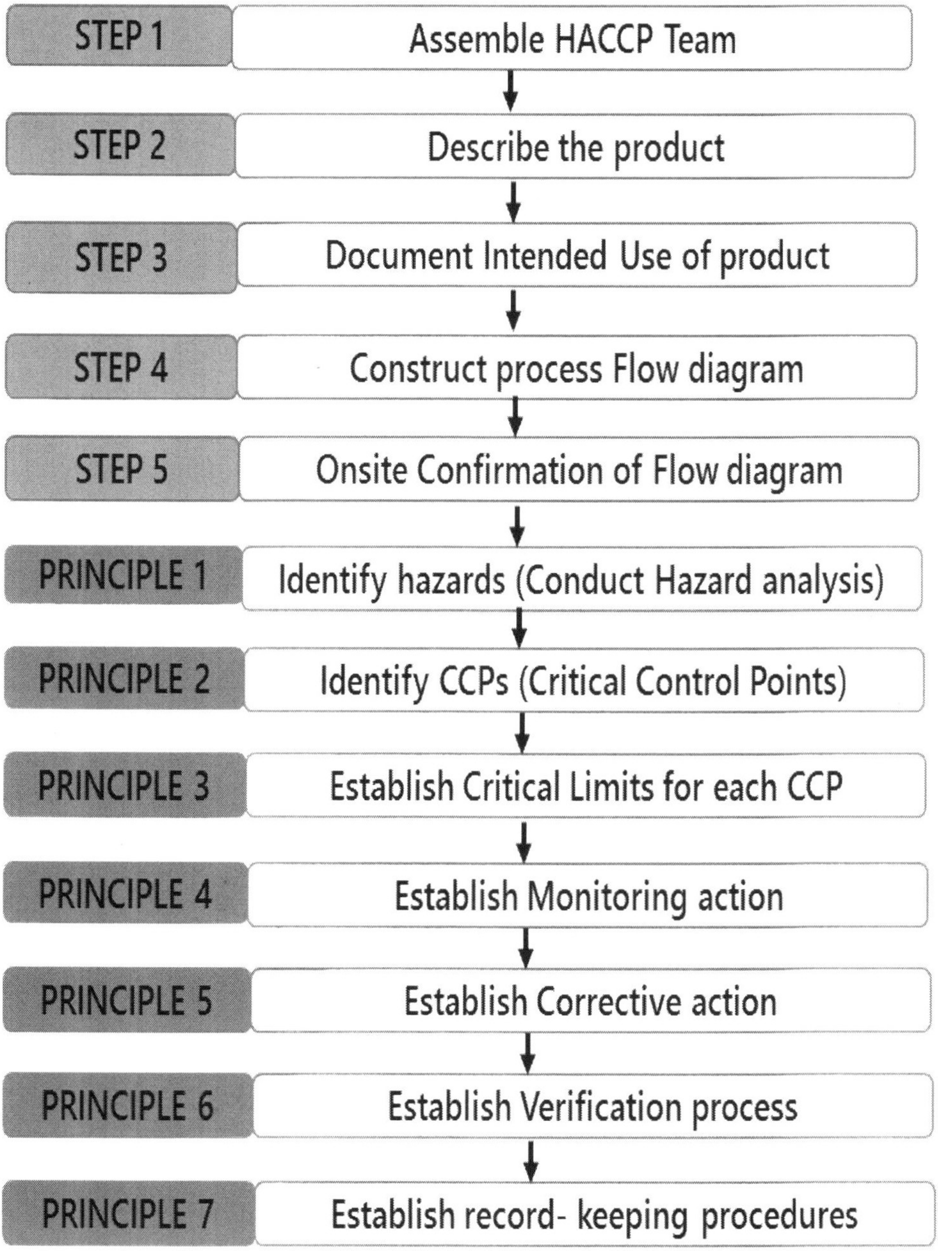

The requirements for Sanitation Standard Operating Procedures (SSOPs) along with Good Manufacturing Practices (GMPs) & Good Hygiene Practices should be considered as Pre- Requisite for HACCP.

Risk assessment is a critical step in a HACCP plan. Below is a template to determine what severity and probability a processing step is involved with and therefore what level of criticality is holds in the processing line.

		Consequence/ Severity				
		How severe could the outcome be if the risk event occurs?				
		Severe	Major	Significant	Minor	Insignificant
Probability/ Likelihood What's the chance of the risk occuring?	Frequent	Extreme	Extreme	Very High	High	Medium
	Likely	Extreme	Very High	High	Medium	Medium
	Occasional	Very High	High	Medium	Medium	Low
	Seldom	High	Medium	Medium	Low	Very Low
	Unlikely	Medium	Medium	Low	Very Low	Very Low

Introduction to Decision Tree

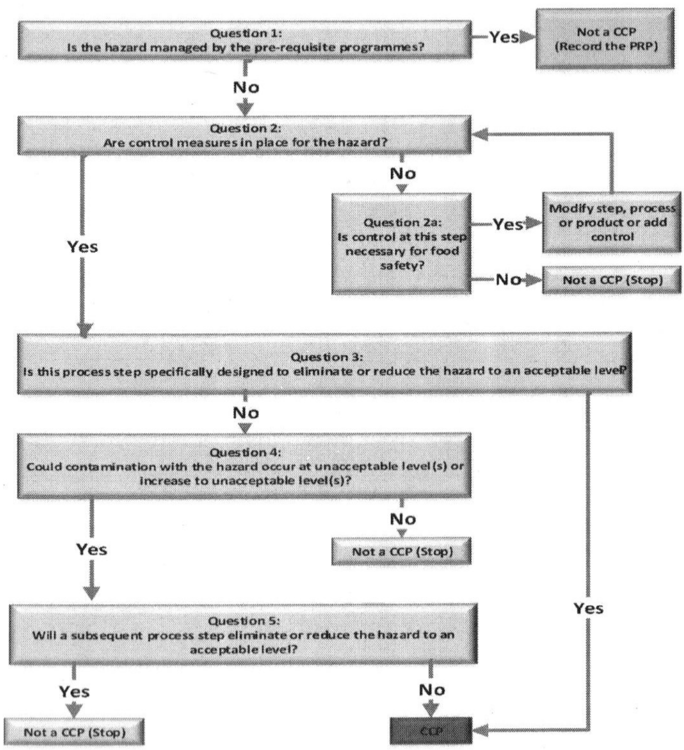

Hazard Analysis and Critical Control Point (HACCP) decision trees are tools that can be used to help you decide whether a hazard control point is a critical control point (CCP) or not. A CCP is a step at which control can be applied. However, it is not always possible to eliminate or prevent a food safety hazard, so this allows you to reduce it to an acceptable level.

The purpose of a decision tree is to support the judgement of the team and help you to confirm whether the hazard needs more food safety controls. Decision trees are not mandatory elements of HACCP but they can be useful in helping you determine whether a particular step is a CCP.

It is vital that you determine the correct CCPs to ensure that food is managed effectively and safely. The number of CCPs in a process will depend on how complex the process is and how many hazards are present.

2. APPLICATION OF HACCP SYSTEM

1. HACCP Implementation steps

1.1 Assemble HACCP team

The food operation shall ensure that the appropriate product specific knowledge and expertise is available for the development and implementation of an effective HACCP plan. A multidisciplinary team shall be assembled either in-house or if such expertise is not available on-site , expert advice shall be obtained from other sources, such as trade and industry associations, independent experts, regulatory authorities, HACCP plan shall be identified and shall describe which segment of the food chain is involved and the general classes of hazards to be addressed (all or selected classes).

1.2 Describe product

A full description of the product shall be drawn up, including relevant safety information such as composition (including raw materials ingredients, allergens), origin,physical/chemical properties that impact food safety (including Aw, pH, etc.),microbial/static treatments (heat treatment, freezing, brining, smoking etc.), packing, labelling, durability and storage conditions and method of distribution. Within businesses with multiple product for example, catering operations with similar characteristics or processing steps may be grouped for the purpose of development of the HACCP plan.

1.3 Identify intended use

The intended use of the product shall be defined based on the expected uses of the product by the end user or customer. The suitability of the product for vulnerable groups of the population such as pregnant women, infants, elderly should be considered, as necessary.

1.4 Construct flow diagram

The flow diagram shall be prepared to cover all steps in the operation for each specific product or product category. When applying HACCP to a given operation, consideration shall be given to steps preceding and following the specified operation.

1.5 On-site confirmation of flow diagram

Steps shall be taken to confirm the proceeding operation against the flow diagram during all stages and hours of operation and amend the flow diagram where appropriate. The confirmation of the flow diagram should be performed by a competent person or persons. On-site verification activities shall be carried out whenever there are any changes in the process.

1.6 List of all potential hazards associated with each step, conduct a hazard analysis, and consider any measures to control identified hazards (SEE PRNCILPLE 1)

The HACCP team should list all potential hazards (physical, chemical, biological) that may be reasonably expected to occur at each step according to the scope.

It should then conduct a hazard analysis to identify for the HACCP plan which hazards are of such a nature that their elimination or reduction to acceptable levels is essential to the production of safe food.

In conducting the hazard analysis, the following should be included as appropriate:

- The likely occurrence of hazard and severity of their adverse health effects;
- The qualitative and/ or quantitative evaluation of the presence of hazards;
- Survival or multiplication of micro-organisms of concern;
- Production of persistence of foods of toxins, chemicals or physical agents; and
- Conditions leading to the above.

For selection of control measures, consideration shall be given to what control measures, if any, can be applied to each hazard.

More than one control measure may be required to control a specific hazard and more than one hazard may be controlled by a specified control, measure. Where elimination of hazard is not practical, justification for acceptable levels of the hazard in the finished product shall be determined and documented.

1.7 Determine Critical Control Points (SEE PRINCIPLE 2)

For each hazard that requires control, control measures shall be identified. The control measures shall be reviewed to identify those that need to be addressed through the HACCP plan and for which CCPs shall be identified. There may be more than one CCP at which control is applied to address the same hazard or there may be cases where there is no CCP identified. The CCP in the HACCP system shall be determined and this may be facilitated by a logic reasoning approach such as the application of a decision tree (see dia 2). The application of a decision tree should be flexible. This example of a decision tree may not be applicable to all situations and alternative approaches may be used.

If a hazard has been identified at a step where control is necessary for safety, and no control measure exists at that step, or any other, then the product or process should be modified at that step, or at any earlier or later stage, to include a control measure.

1.8 Establish Critical Limits for each CCP (SEEPRINCIPLE 3)

Critical Limits shall be specified and validated for each CCP. In some cases more than one critical limit may be elaborated at a particular step.

These critical limits shall be measurable, Critical Limits based on subjective data (such as visual inspection of product, process, handling) shall be supported by instructions or specifications and / or education and training.

1.9 Establish a monitoring system for each CCP (SEE PRINCIPLE 4)

A monitoring system shall be established for each CCP to demonstrate that the CCP is under control. The monitoring shall be able to detect loss of control at the CCP and in time to make adjustments to regain control of the process and prevent violation of the critical limits. Where possible, process adjustments should be made when the results of monitoring indicate a trend towards loss of control at a CCP. The adjustment should be taken before a deviation occurs.

Data derived from monitoring shall be evaluated by a designated person with knowledge and authority to carry out corrective actions when indicated. If monitoring is not continuous, then the amount or frequency of monitoring shall be sufficient to ensure that the CCP is under control. The monitoring system shall cover the following:

a) Measurements or observations that provide results within an adequate time frame;
b) Monitoring device used;
c) Applicable calibration method;
d) Monitoring frequency;
e) Responsibility and authority related to monitoring and evaluation of monitoring results; and
f) Records.

All records and documents associated with monitoring CCPs shall be signed by the person(s) doing the monitoring and by the responsible reviewing official(s) of the company.

The monitoring methods and frequency shall be capable of determining when the critical limits have been exceeded in time for the product to be isolated before it is used or consumed.

1.10 Establish corrective actions (SEE PRINCIPLE 5)

Specific planned corrective actions shall be developed for each CCP in the HACCP system in order to deal with deviations when they occur and to prevent their recurrence. This may require identification of the causes of deviation.

The action shall ensure that the CCP has been brought under control. Actions taken shall also include proper disposition of the affected product. Deviation and product disposition procedures shall be documented. Records of deviations and disposition shall be maintained.

1.11 Establish Verification Procedures (SEE PRINCIPLE 6)

The verification procedures consist of two activities, verification activities and validation activities.

The food business operator shall have in place a system to verify the HACCP plan at a set frequency. Procedures for verification shall be established. The frequency of verification should be sufficient to confirm that the HACCP system is working effectively.

Verification should be carried out by someone other than the person who is responsible for performing the monitoring and corrective actions. Where certain verification activities cannot be performed in-house, verification should be performed on behalf of the business by external experts or qualified third parties.

The HACCP system, including the HACCP plan, shall be reviewed (atleast once in a year) and necessary changes made when any modification is made in the product, process, or any step.

- Verification activities shall include:
- Self-evaluation;
- Review of the HACCP system and plan and its records;
- Review of deviation and product dispositions; and
- Confirmation that CCPs are kept under control.

The results of verification shall be maintained and communicated to the HACCP team/ relevant staff.

The food business operator shall periodically validate the HACCP plan and necessarily before its implementation and after any changes are made. The objective of the validation process is to ensure that identified hazards are complete, correct and effectively controlled under the HACCP plan. Validation activities should include actions to confirm the efficacy of the HACCP system. Records of validation shall be maintained. An annual review of the complete HACCP system shall be carried out.

Verification and validation activities are also important for maintenance of the system as well as continual improvements.

1.12 Establish Documentation and Record Keeping (SEE PRINCIPLE 7)

HACCP procedures shall be documented. Documentation and record keeping shall be appropriate to the nature and size of the operation and sufficient to assist the business to verify that the HACCP controls are in place and being maintained.

Documentation shall include (as a minimum) the following:

- HACCP team composition;
- Product description;
- Intended use;
- Flow chart;
- Hazard analysis;
- CCP determination;
- Critical limit determination;
- Validation process; and
- HACCP plan

The HACCP plan shall include the following information for each identified CCP:

- Food safety hazard(s) to be controlled at the CCP;
- Control measure(s);
- Critical limit(s);
- Monitoring procedure(s);
- Corrections and corrective action(s) to be taken if critical limits are exceeded;

- Responsibilities and authorities for monitoring, corrective action and verification;
- Record(s) of monitoring.
- Records to include
- CCP monitoring activities;
- Deviations and associated corrective actions;
- Disposition of non-conforming products;
- Verification procedures performed;
- Modifications to the HACCP plan;
- Validation record;
- Product release records; and
- Testing records.

2.1 Process Flow Chart

2.1.1 Biscuits

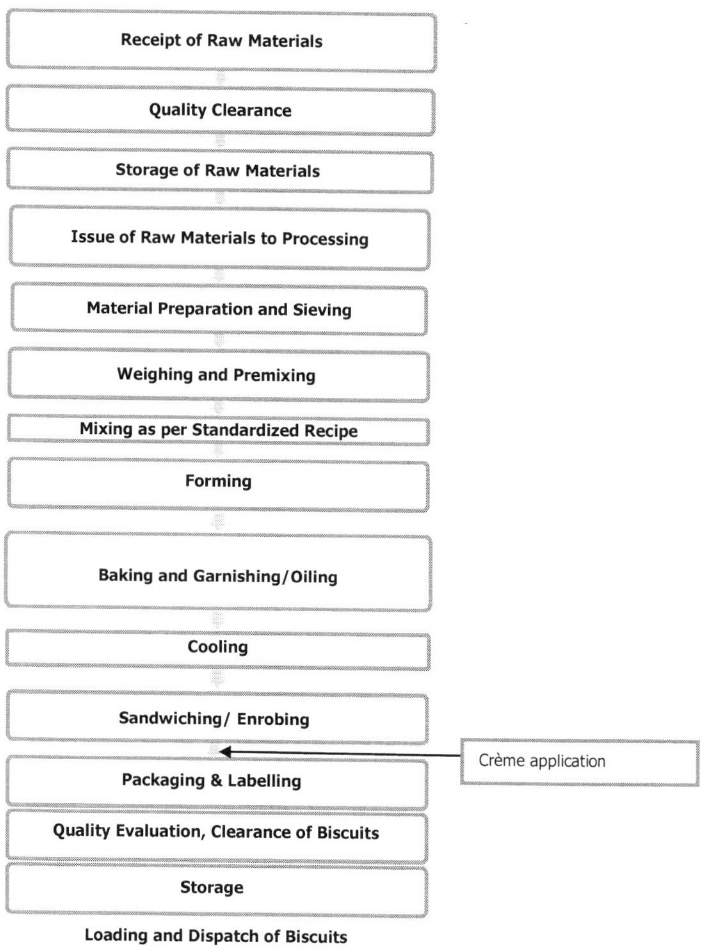

2.1.2 Pie

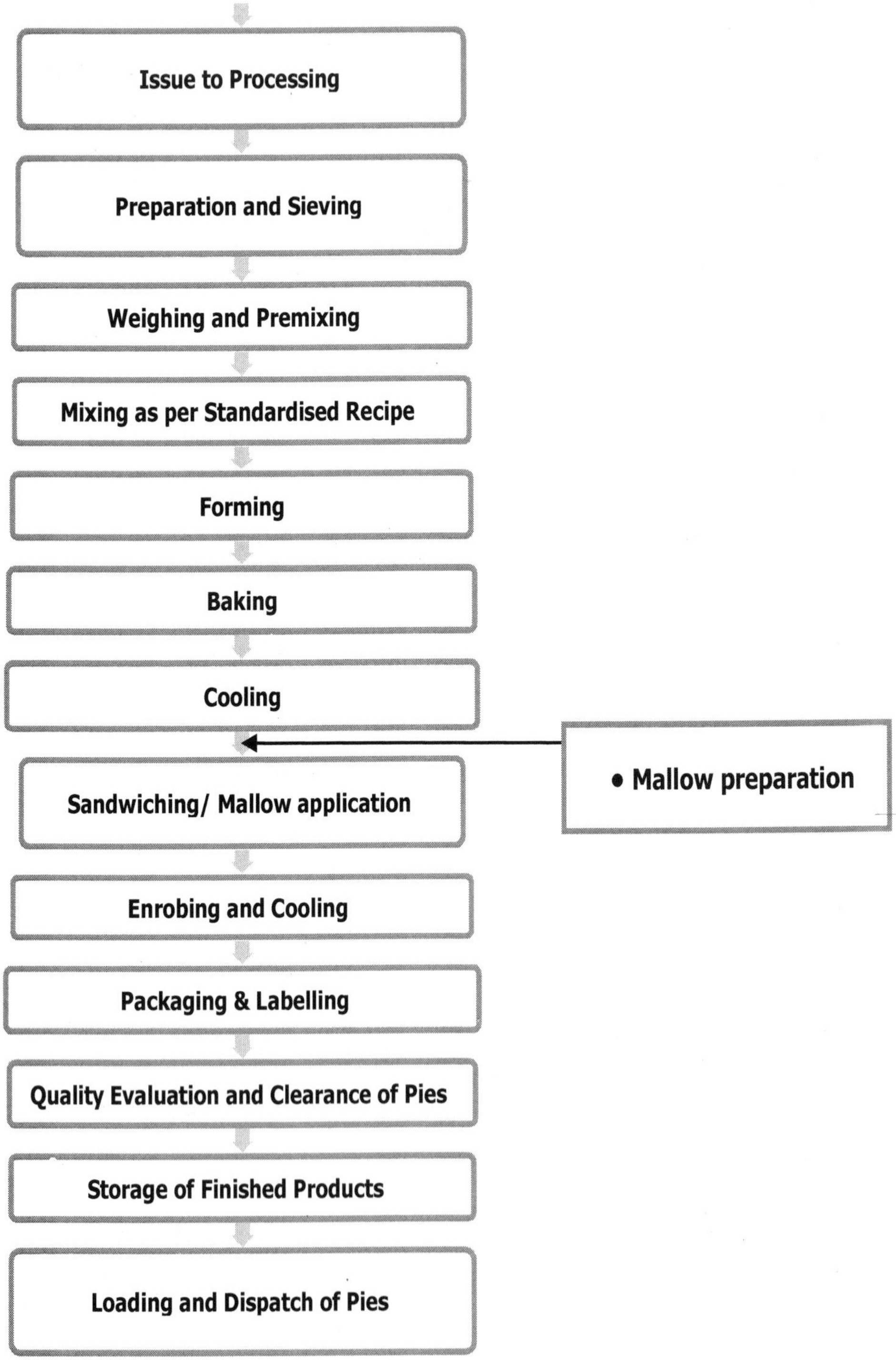
Issue to Processing
Preparation and Sieving
Weighing and Premixing
Mixing as per Standardised Recipe
Forming
Baking
Cooling
• Mallow preparation
Sandwiching/ Mallow application
Enrobing and Cooling
Packaging & Labelling
Quality Evaluation and Clearance of Pies
Storage of Finished Products
Loading and Dispatch of Pies

2.2 Hazard Analysis & Identification for Biscuits & Pies

Sl. No.	List of Manufacturing/ Process Steps / (sequential)	Possible Physical Hazards	Possible Chemical Hazards	Possible Biological Hazards	Possible Allergens	Control Measures
1.	**Receiving ingredients:**					
	Dry material in gunny bags or plastic containers	Foreign bodies like Metal, Jute, Thread, Hair, Stone, Glass, Cement, Black Particles, Rice Husk	Vehicle grease	Weevils, Larvae, Insects	Nuts, Eggs, Gluten	Conduct inspection, effective Allergen management plan, periodic testing
	Liquid in drums	Wood, Plastic			Milk	Conduct inspection
2.	**Ingredient Storage:**					
	Cocoa Powder, Butter, Flour, Essences, Egg	Foreign bodies like Metal, Jute, Thread, Hair, Stone, Glass, Cement, Black Particles	Cross contaminatio n cleaning or non-food chemicals	Rodents, pests and insects, fungal growth in case of damp storage environment, fermentation	Nuts, Eggs, Gluten and Milk	Maintain hygienic storage condition, effective Allergen management plan
3.	**Pre-Mixing:**					
	Flour sifting	Thread, Hair, Stone, Glass, Plastic,	Weevil, Larvae		Gluten	Sifting of flour, sieve integrity checks, magnet strength check, periodic testing
	Sugar passing through magnetic grill	Metal etc				Magnetic grill

Sl. No.	List of Manufacturing/ Process Steps / (sequential)	Possible Physical Hazards	Possible Chemical Hazards	Possible Biological Hazards	Possible Allergens	Control Measures
	Egg breaking	Egg shell		Pathogens like Salmonella etc	Egg	Addition of Egg whisk through egg shell sieve, COA of vendor
	Preparation of colour solution		Colour concentration beyond acceptable limit			Check for acceptable limit
	Water testing	Sand	All heavy metals, pesticides and radioactive substances	Coliforms, E Coli		Periodic water quality testing and RO/ UV system singularly or in combination
4.	**Mixing of ingredients:**					
	Addition of materials in mixer	Thread, Hair, Stone, Glass, Plastic	Grease		Nuts, Milk, wheat flour, soya lecithin	Sifting of all ingredients and Cleaning of all parts of mixer, effective Allergen management plan
	Addition of Egg whisk	Egg shell		Pathogens like Salmonella etc	Egg	Addition of Egg whisk through egg shell sieve, COA of vendor
5.	**Baking:**					

Sl. No.	List of Manufacturing/ Process Steps / (sequential)	Possible Physical Hazards	Possible Chemical Hazards	Possible Biological Hazards	Possible Allergens	Control Measures
		Metal	Grease	Microbial contamination		Passing through metal detectors, Baking at proper temperature and time to avoid microbial growth, periodic moisture check to control water activity and regular cleaning of the oven
6.	**Cooling:**					
				Microbial contamination, pest and insect infestation		PRP and regular air quality check, Infestation Control programs
7.	**Cream/ Mallow** Preparation and application and/ or Enrobing	Hair, Metal and other Foreign matters	Cleaning chemical residues	Yeast and Mold, E Coli	Milk, Soya	Metal detector for sandwiches, GHP controls, sieves, cleaning programs, Vendor COA, Microbial testing for FG, Allergen management plan
8.	**Packing:**					

Sl. No.	List of Manufacturing/ Process Steps / (sequential)	Possible Physical Hazards	Possible Chemical Hazards	Possible Biological Hazards	Possible Allergens	Control Measures
			Grease, solvent odour	Infestation, Aerial contamination		Usage of Food Grade lubricants on Food contact surface, seal integrity control, vendors certificate C0A on overall migration, infestation control program
9.	**Storing:**					
			Cross contamination from Non-food chemicals			Maintain hygienic storage condition.
10.	Loading and Dispatch:		Grease			Inspection of vehicle

2.3 HACCP Implementation Plan for Biscuits & Pies

Step	Hazards	Control Measures / CCP	Critical Limit	Monitoring			Cor- rec- tive action	Verif- ic- ation Responsibi lity	Reco- rds
				Proced ure	Frequ- ency	Responsiility			
Ingredient Receiving	Foreign bodies like Metal, Jute, Stone, Glass, Cement	Carrier Inspection, Sieving and Mag- net	Nil	Visual Checki ng and integra ted control s	Every con- sign ment	Stores Super- visor	Reject, if not found OK	QA team	Incomi ng check records
Ingredient Storage	Off odour, infesta- tion And fermentati On	Physical inspection, Counts on traps	Absent	Monit oring of Cold room tem- pe rature Infesta tion control progra m, ven- dor declar ation, periodic fumiga tion of strateg ist locatio ns	Every 24 hours temp check, once in 3 months or internally defined fumigati on plan	Stores Supervi- sor, Pest Con- trol Agency	Reject or re- fumi- gate, if not found OK	QA team	Phys- ica l Inspecti on and Storage Temper ature, Pest Control Records
Pre-Mixing	Thread, Weevils, Larvae	Physical inspectio n through sieving in BSS 36 mesh or internally defined size for in- gredien ts	Absent	Checki ng con- diti on of sieve and use of prop- er sieve size	Every 24 hours or internally defined frequenc y	Production Supervisor	Immedia te replace ment of damage dsieve,. Checking of sieved material Proper dispos- al of material collected on sieve	QA team	Sieve checkli st

Step	Hazards	Control Measures / CCP	Critical Limit	Monitoring			Cor- rec- tive action	Verif- ic- ation Responsibi lity	Reco- rds
				Proced ure	Frequ- ency	Responsiility			
	Colour Beyond acceptabl	Assuranc e of colour	As per recipe	Weight checki ng for approp	Once in a week	Production Supervisor	Rejectio n or dilution of colour	QA team	Colour Prepar Ation
	e limit	quantity		riate quantit y			solution		Record
Mixing	Hair, Thread etc	Sieving of material	Nil	Checki ng of sieve	Every 24 hours	Production Supervisor	Replace ment of damage d sieve and recheck- ing of sieved materials	QA Team	Sieve checkli st
	Egg shell	Sieving of Egg	Nil	Checki ng of sieve	Every 24 hours	Production Supervisor	Replace ment of damage d sieve and rechecki ng of sieved materials	QA Team	Sieve checkli st
Baking	Microbial contamina tion	Proper maintena nce of baking time and temperat ure	Nil	Checki ng of Baking time and tempe rature, moistu re check	Twice in each shift or internally defined frequenc y	Production Supervisor or Line Inspector	Stoppag e of Producti on and rejection of producti on	QA Team	Baking time, temp, moistur e records

Step	Hazards	Control Measures / CCP	Critical Limit	Monitoring			Cor- rec- tive action	Verif- ic- ation Responsibi lity	Reco- rds
				Proced ure	Frequ- ency	Responsiility			
Cooling	Microbial contamina Tion	Microbial count	Within accepta ble limit	Daily air sampling, ozone dosage check, UV interlo ck checks	Once in 24 hours or internally defined frequency	QA Team	Stop- page of Produc- tion and thorough fumiga- tion	QA Manager	Microbi Ology Records
Cream/ Mallow Preparatio n and applica- tion and/ or Enrobing	Physi- cal and microbial contamina tion	GHP plan, Lint rollers, Metal De- tector Yeast and Mold, E Coli	Nil Within specifie d limits	GHP checkli st, MD verifica tionMi crobial testing	As per internally defined frequenc y	QA Team		QA Man- ager	GHP, CCP records, Microbi ology records
Packing	Aerial contamina tion due to Improper packing	Air counts	As per defined internal limits	Microb ial testing	Every day on line	Packing Super- visor	Rejectio n of Prod- uct	QA Team	Microbi al lab records
	Improper sealing	Physical observati on	Absent	Visual inspect ion/Va cuum leak test	Every shift on line	Packing Super- visor	Packing machine to be stopped and recti- fied .Recy- clin g of defectiv e packets	QA Team	Packing machin e VLT records

3. HACCP Plan for Bread

3.1 Process Flow Chart for Bread

- **Receipt of Raw Materials**
- **Quality Clearance**
- **Storage of Raw Materials**
- **Issue to Processing**
- **Material Sieving**
- **Preparation as per Standardised Recipe**
- **First Mixing**
- **Bulk fermentation (BFT process)**
- **Second Mixing**
- **Dividing and Rounding**
- **Interproofing**
- **Moulding**
- **Final Proofing**
- **Baking**
- **Depanning**
- **Cooling**
- **Slicing and Packaging**
- **Quality Evaluation, Clearance of Bread**
- **Storage of Finished Bread**
- **Loading and Dispatch of Breads**

3.2 Hazard Analysis & Identification for Bread

S. No.	List of Manufacturing/ Process Steps / (sequential)	Possible Physical Hazards	Possible Chemical Hazards	Possible Biological Hazards	Possible Allergens	Control Measures
1.	**Receiving ingredients:**					
	Dry in gunny bags or plastic containers	Foreign bodies like Metal, Jute, Thread, Hair, Stone, Glass, Cement, Black Particles, Rice Bran	Vehicle grease	Weevils, Larvae, Insects	Wheat	Conduct inspection
	Liquid in drums	Wood, Plastic			Milk	Conduct inspection
2.	**Ingredient Storage**:					
	Butter, Flour, Essences	Unacceptable odour due to spoilage or fermentation	Cross contamination cleaning or non-food chemicals chemicals	Rodents, pests and insects, fungal growth in case of damp storage environment		Maintain hygienic storage condition.
3.	Pre-Mixing					
	Flour sifting	Thread, Hair, Stone, Glass, Plastic		Weevil, Larvae	Wheat flour	Sifting of flour
	Water testing		Arsenic			Periodic water quality testing

S. No.	List of Manufacturing/ Process Steps / (sequential)	Possible Physical Hazards	Possible Chemical Hazards	Possible Biological Hazards	Possible Allergens	Control Measures
4.	**Mixing of ingredients**					
	Addition of materials in mixer	Thread, Hair, Stone, Glass, Plastic	Grease		Milk, wheat flour	Sifting of all ingredients and Cleaning of all parts of mixer.
		Dirt, Dust				Periodic cleaning of batter rollers
5.	**Proofing**					
		Dirt, Dust		Insects and pests		Clean prooving area regularly
6.	**Baking**					
		Metal	Grease	Microbial contamination		Passing through metal detectors, Baking at proper temperature and time to avoid microbial growth and Regular cleaning of the oven
7.	**Cooling**					
				Microbial contamination, pest and insect infestation		PRP and regular air quality check

S. No.	List of Manufacturing/ Process Steps / (sequential)	Possible Physical Hazards	Possible Chemical Hazards	Possible Biological Hazards	Possible Allergens	Control Measures
8.	**Slicing**					
		Dust and Dirt				Periodic cleaning of slicer blades and belts
9.	**Packing**					
				Aerial contamination		
10.	**Storing**					
			Cross contamination from Non-food chemicals			Maintain hygienic storage condition.
11.	**Loading and Dispatch**					
			Grease			Inspection of vehicle

3.3 HACCP Implementation Plan for Bread

Step	Hazards	Control Measures / CCP	Critical Limit	Monitoring			Corrective action	Verification Responsibi lity	Records
				Procedure	Frequency	Responsibility			
Ingredients Receiving	Foreign bodies like Metal, Jute, Stone, Glass, Cement, etc;	Carrier Inspection	Nil	Visual Checki ng	Every consignme nt	Stores Supervisor	Reject, if not found OK	QA team	
Ingredient Storage	Off odour	Physical inspectio n	Absent	Monit oring of Cold room tempe rature and Shelf life	Every 24 hours	Stores Supervisor	Reject, if not found OK	QA team	
Pre-Mixing	Thread, Weevils, Larvae	Physical inspectio n through sieving in 35 mesh	Absent	Checki ng of sieve and use of proper sieve size	Every 24 hours	Production Supervisor	1.Imme diate replace ment of damag ed sieve, and checkin g of sieved materia l 2. Proper disposa l of materia l collecte d on sieve	QA team	
Mixing	Hair, Thread etc	Sieving of material	Nil	Checki ng of sieve	Every 24 hours	Production Supervisor	Replac ement of damag ed sieve and recheck	QA Team	

4. HACCP Plan for Cakes

4.1 Process Flow Chart for Cakes

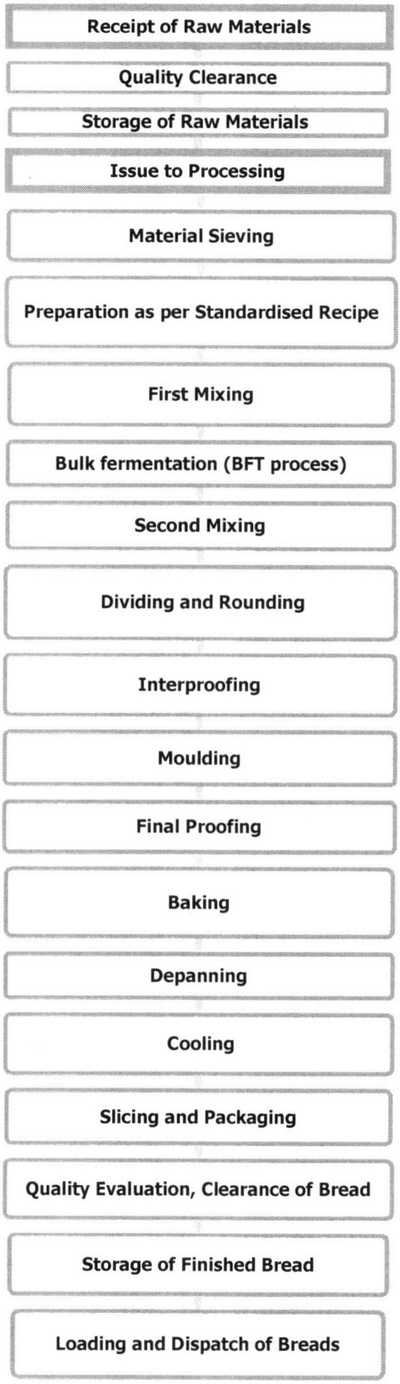

4.2 Hazard Analysis & Identification for Cakes

S. No.	List of Manufacturing/ Process Steps / (sequential)	Possible Physical Hazards	Possible Chemical Hazards	Possible Biological Hazards	Possible Allergens	Control Measures
1.	Receiving ingredients:					
	Dry in gunny bags or plastic containers	Foreign bodies like Metal, Jute, Thread, Hair, Stone, Glass, Cement, Black Particles, Rice Bran	Vehicle grease	Weevils, Larvae, Insects	Nuts, Eggs, Wheat	Conduct inspection
	Liquid in drums	Wood, Plastic			Milk	Conduct inspection
	Fruits that are cut	Seeds				Visual Inspection
2.	Ingredient Storage:					
	Cocoa Powder, Butter, Flour, Essences	Unacceptable odour due to spoilage or fermentation	Cross contaminatio n cleaning or non-food chemicals chemicals	Rodents, pests and insects, fungal growth in case of damp storage environment	Nuts,eggs, Wheat and Milk	Maintain hygienic storage condition.
3.	Pre-Mixing					
	Flour sifting	Thread, Hair, Stone, Glass, Plastic		Weevil, Larvae	Wheat flour	Sifting of flour
	Sugar passing through magnetic grill	Metal				Magnetic grill
	Egg breaking	Egg shell			Egg	Addition of Egg whisk through egg shell siever

S. No.	List of Manufacturing/ Process Steps / (sequential)	Possible Physical Hazards	Possible Chemical Hazards	Possible Biological Hazards	Possible Allergens	Control Measures
	Preparation of colour solution		Colour concentration beyond acceptable limit			Check for acceptable limit
	Water testing		Arsenic			Periodic water quality testing
4.	Mixing of ingredients					
	Addition of materials in mixer	Thread, Hair, Stone, Glass, Plastic	Grease		Nuts, Milk, wheat flour	Sifting of all ingredients and Cleaning of all parts of mixer.
	Addition of Egg whisk	Egg shell			Egg	Addition of Egg whisk through egg shell siever
5	Batter scaling and levelling					
		Dirt, Dust				Periodic cleaning of batter rollers
6.	Baking					
		Metal	Grease	Microbial contamination		Passing through metal detectors, Baking at proper temperature and time to avoid microbial growth and Regular
						cleaning of the oven
7.	Cooling					

S. No.	List of Manufacturing/ Process Steps / (sequential)	Possible Physical Hazards	Possible Chemical Hazards	Possible Biological Hazards	Possible Allergens	Control Measures
				Microbial contamination, pest and insect infestation		PRP and regular air quality check
8.	Slicing					
		Dust and Dirt				Periodic cleaning of slicer blades and belts
9.	Packing					
				Aerial contamination		
10.	Storing					
						Maintain hygienic storage condition.
11.	Loading and Dispatch					
						Inspection of vehicle

2.3 HACCP Implementation Plan for Cakes

Tep	Hazards	Control Measures /CCP	Critical Limit	Monitoring			Corrective action	Verification Responsibility	Recor ds
				Proced ure	Frequency	Responsibility			
Ingredient s Receiving	Foreign bodies like Metal, Jute, Stone, Glass, Cement, etc;	Carrier Inspectio n	Nil	Visual Checkin g	Every consign ment	Stores Supervisor	Reject, if not found OK	QA team	
Ingredient Storage	Off odour	Physical inspectio n	Absent	Monitor ing of Cold room temper ature and Shelf life	Every 24 hours	Stores Supervisor	Reject, if not found OK	QA team	
Pre- Mixing	Thread, Weevils, Larvae	Physical inspectio n through sieving in 35 mesh	Absent	Checkin g of sieve and use of proper sieve size	Every 24 hours	Production Supervisor	1. Immediate replacement of damaged sieve, and checking of sieved material 2. Proper disposal of material collected on sieve	QA team	
	Colour Beyond acceptable limit	Assurance of colour quantity	As per recipe	Weight checking for appropriate quantity	Once in a week	Production Supervisor	Rejection or dilution of colour solution	QA team	
Mixing	Hair, Thread etc	Sieving of material	Nil	Checkin g of sieve	Every 24 hours	Production Supervisor	Replacemen t of damaged sieve and rechecking of sieved materials	QA Team	

Tep	Hazards	Control Measures /CCP	Critical Limit	Monitoring			Corrective action	Verification Responsibility	Recor ds
				Proced ure	Frequency	Responsibility			
	Egg shell	Sieving of Egg	Nil	Checking of sieve	Every 24 hours	Production Supervisor	Replacement of damaged sieve and rechecking of sieved materials	QA Team	
Baking	Microbial contamination	Proper maintena nce of baking time and temperat ure	Nil	Checkin g of Baking time and temper ature	Twice in each shift	Production Supervisor	Stoppage of Production and rejection of production	QA Team	
Cooling	Microbial contamination	Microbial count	Within accept able limit	Daily air sampling	Once in 24 hours	QA Manager	Stoppage of Production and thorough fumigation	QA Team	
Slicing	Dirt and dust, Microbial contamination	Swab test	Absent	Swab test of slicing blades	Thrice in a week	QA Manager	Ensure proper cleaning	QA Team	
Packing	Aerial contamina tion due to Improper packing	Physical observation	Absent	Visual inspecti on	Every day on line	Packing Supervisor	Rejection of Product	QA Team	
	Improper sealing	Physical observation	Absent	Visual inspecti on	Every day on line	Packing Supervisor	Packing machine to be stopped and rectified Recycling of defective packets	QA Team	

5. History of HACCP

HACCP was developed by the Pillsbury Company while working on producing foods for NASA for use in space missions in early 1959. NASA had concerns of food, particularly crumbs, in the space capsule in zero gravity and also food that was free of pathogens and biological toxins that Pillsbury addressed by the use of HACCP.

The concept of HACCP was first presented to the public in the 1971 National Conference on Food Protection. At that time it was based on three principles.

In 1985, interest in HACCP was renewed when a subcommittee of the Food Protection Committee of NASA issued a report on microbiological criteria.

A National Advisory Committee on Microbiological Criteria for Foods was formed and that committee published a report in 1992 that provided the framework for HACCP as we know it today.

6. Principles of HACCP

The report by the National Advisory Committee on Microbiological Criteria for Food listed the Seven Principles of HACCP:

Principle 1 - Conduct A Hazard Analysis

The application of this principle involves listing the steps in the process and identifying where significant hazards are likely to Occur.

The HACCP team will focus on hazards that can be prevented, eliminated or controlled by the HACCP plan.

A justification for including or excluding the hazard is reported and possible control measures are identified.

Principle 2 - Identify the Critical Control Points

A critical control point (CCP) is a point, step or procedure at which control can be applied and a food safety hazard can be prevented, eliminated or reduced to acceptable levels.

The HACCP team will use a CCP decision tree to help identify the critical control points in the process.

A critical control point may control more than one food safety hazard or in some cases more than one CCP is needed to control a single hazard.

The number of CCP's needed depends on the processing steps and the control needed to assure food safety.

Principle 3 - Establish Critical Limits

A critical limit (CL) is the maximum and/or minimum value to which a biological, chemical, or physical parameter must be controlled at a CCP to prevent, eliminate, or reduce to an acceptable level the occurrence of a food safety hazard.

The critical limit is usually a measure such as time, temperature, water activity (Aw), pH, weight, or some other measure that is based on scientific literature and/ or regulatory standards.

Principle 4-Monitor CCP

The HACCP team will describe monitoring procedures for the measurement of the critical limit at each critical control point.

Monitoring procedures should describe how the measurement will be taken, when the measurement is taken, who is responsible for the measurement and how frequently the measurement is taken during production.

Principle 5 - Establish Corrective Action

Corrective actions are the procedures that are followed when a deviation in a critical limit occurs.

The HACCP team will identify the steps that will be taken to prevent potentially hazardous food from entering the food chain and the steps that are needed to correct the process.

This usually includes identification of the problems and the steps taken to assure that the problem will not occur again.

Principle 6 - Verification

Those activities, other than monitoring, that determine the validity of the HACCP plan and that the system is operating according to the plan.

The HACCP team may identify activities such as auditing of CCP's, record review, prior shipment review, instrument calibration and product testing as part of the verification activities.

Principle 7 - Recordkeeping

A key component of the HACCP plan is recording information that can be used to prove that food was produced safely.

The records also need to include information about the HACCP plan.

Record should include information on the HACCP Team, product description, flow diagrams, the hazard analysis, the CCP's identified, Critical Limits, Monitoring System, Corrective Actions, Recordkeeping Procedures, and Verification Procedures.

FOOD SAFETY AND STANDARDS AUTHORITY OF INDIA (FSSAI)

fssai

FOOD SAFETY AND STANDARDS
AUTHORITY OF INDIA

Food Safety and Standards Authority of India (FSSAI) has been established under Food Safety and Standards, 2006 to handle food related issues in various Ministries and Departments. FSSAI has been created for laying down science-based standards for articles of food and to regulate their manufacture, storage, distribution, sale and import to ensure availability of safe and wholesome food for human consumption. One of the functions of FSSAI is to maintain the quality of food in the country.

Introduction

FSSAI provides an opportunity for Food Business Operators (FBOs) to apply for Registration/Licensing through online Registration/Licensing System. It also provides an option to track the status of application online through an Application Reference No. provided by the system during the submission of application.

The FBOs should take a print out of the Online Application Form generated through the system and submit the application to Regional Authority/State Authority with all supportive documents within fifteen days from the date of online submission of application for State License and Registration Certificate. In case State License and Registration Certificate of Delhi and for Central License all documents are to be uploaded electronically and no physical documents are to be submitted in the Regional Office

FBO can register themselves online for the username and password (or) by providing the contact details can apply for Licensing / Registration.

The Important Functions of FSSAI:

- Framing of Regulations to lay down the Standards and guidelines in relation to articles of food and specifying appropriate system of enforcing various standards thus notified.
- Laying down mechanisms and guidelines for accreditation of certification bodies engaged in certification of food safety management system for food businesses.
- Laying down procedure and guidelines for accreditation of laboratories and notification of the accredited laboratories.
- To provide scientific advice and technical support to Central Government and State Governments in the matters of framing the policy and rules in areas which have a direct or indirect bearing of food safety and nutrition.
- Collect and collate data regarding food consumption, incidence and prevalence of biological risk, contaminants in food, residues of various, contaminants in foods products, identification of emerging risks and introduction of rapid alert system.
- Creating an information network across the country so that the public, consumers, Panchayats etc receive rapid, reliable and objective information about food safety and issues of concern.
- Provide training programmes for persons who are involved or intend to get involved in food businesses.

- Contribute to the development of international technical standards for food, sanitary and phyto-sanitary standards.
- Promote general awareness about food safety and food standards.

I. FOOD SAFETY AND STANDARDS REGULATIONS

1. Food Safety and Standards (Licensing and Registration of Food Businesses) Regulation, 2011
2. Food Safety and Standards (Food Products Standards and Food Additives) Regulation, 2011
3. Food Safety and Standards (Prohibition and Restriction of Sales) Regulation, 2011
4. Food Safety and Standards (Packaging and Labelling) Regulation, 2011
5. Food Safety and Standards (Contaminants, Toxins and Residues) Regulation, 2011
6. Food Safety and Standards (Laboratory and Sampling Analysis) Regulation, 2011
7. Food Safety and Standards (Health Supplements, Nutraceuticals, Food for Special Dietary Use, Food for Special Medical Purpose, Functional Food and Novel Food) Regulations, 2016
8. Food Safety and Standards (Food Recall Procedure) Regulation, 2017
9. Food Safety and Standards (Import) Regulation, 2017
10. Food Safety and Standards (Approval for Non-Specific Food and Food Ingredients) Regulation, 2017
11. Food Safety and Standards (Organic Food) Regulation, 2017
12. Food Safety and Standards (Alcoholic Beverages) Regulation, 2018
13. Food Safety and Standards (Fortification of Food) Regulation, 2018
14. Food Safety and Standards (Food Safety Auditing) Regulation, 2018
15. Food Safety and Standards (Recognition and Notification of Laboratories) Regulation, 2018
16. Food Safety and Standards (Advertising and Claims) Regulation, 2018
17. Food Safety and Standards (Packaging) Regulation, 2018
18. Food Safety and Standards (Recovery and Distribution of Surplus food) Regulation, 2019

Compliance steps

This act aims to establish a single reference point for all the matters relating to the food safety and the standards by moving from the multilevel, multi-department control to a single line of the command. To this effect, the act establishes an independent statutory authority which is FSSAI whose headquarter is in New Delhi.

There are a lot of steps that have been included in the FSSAI act, 2006 but the main and the mandatory step that needs to be done by the businessman who enters the food business as they have to first register their food company under the FSSAI act, 2006 with FSSAI and then they have to take the license from the FSSAI. It is the mandatory step that all company should follow at the time when they are entering the food market.

There are certain key points that are fulfilled by the companies at the time of the registration. The key points are as follows:-

1. **Basic Entity Registration:** The food business operator must be a registered entity having a separate PAN, VAT registration/ **service tax registration**.
2. **Food License:** It is mandatory for the startup or the company which is planning to enter the food market to get a **food license** from the FSSAI.
3. **Search for the category of license:** Food license is location specific and type of license required varies as per the nature of the activity and the area of the location.
4. **Expedite Processing:** Once the application is filled in the prescribed format along with supporting documents the license can be obtained within 15-20days of the said filling.
5. **The condition of the license:** Food license issued is subject to the certain conditions. All food business operators must ensure that the condition is fulfilled within the time period.
6. **Labeling Requirement:** Logo and license no hall to be displayed on the product package in contrast to the background color as per the labeling regulation.

Food license basically is a permission granted by the FSSAI to carry on the food business. This food license is location specific and is also termed bound. It carries a 14 digit registration number.

There are certain rules and steps that all the companies follow at the time of entering into the food business. The steps for the food license are-

1. Eligibility search.
2. Application Filling.
3. Generation of application id.
4. Scrutiny of application.
5. Issuance of license.
6. Inspection of premises.

The compliance steps included in the FSSAI act are really appreciable and because of these steps, only the humans are able to get safe and healthy food to eat which also helps them to be safe from the diseases. The FSSAI plays a huge role in the fulfillment of the steps and it is the one which makes sure that every company who is entering the food business have fulfilled the steps.

DISPOSAL OF GARBAGE

Waste Disposal

The primary waste in the bread kitchen unit is of unused mixture, fondant, consumed item. These should be evacuated in an ideal way. Batter and fondant may prompt parasitic development in the assembling premises, though consumed item will pollute the completed item.

Waste ought to be arranged promptly in secured containers to evade sullying.

Food squander, non-palatable side-effect and other deny will not be permitted to amass in food dealing with or capacity regions. It will be expelled intermittently with a base every day expulsion in order to stay away from amassing and flood in food taking care of, food stockpiling, other working regions and abutting condition.

Marked materials, items or printed bundling assigned as waste will be deformed or obliterated to guarantee that brand names can't be reused. Expulsion and devastation ought to be done by endorsed removal temporary workers. The association ought to hold records of demolition.

Removal of sewage and effluents (strong, fluid and gas) will be done in congruity with indicated necessities of processing plant act/state contamination control board.

Squander stores and residue receptacles must be kept fittingly spotless, liberated from bugs and in shut conditions and will be arranged according to neighbourhood rules and guidelines including those for plastic and other non-condition inviting materials.

Swachh Bharat Mission

Swachh Bharat Abhiyan (SBA) or Clean India Mission was a nationwide crusade from 2014 to 2019, to dispose of open poo and improve strong waste administration (SWM) in urban and provincial regions in India. The goals of the crucial included destruction of manual searching, producing mindfulness and achieving a conduct change in regards to disinfection practices, and increase of limit at the nearby level. Started by the Government of India, the mission planned to accomplish an "open-poop free" (ODF) India by 2 October 2019, the 150th commemoration of the introduction of Mahatma Gandhi. The mission planned for advancing towards target 6.2 of the Sustainable Development Goals Number 6 set up by the United Nations in 2015.

1. Recycling and Composting Food Waste

Compost

Treating the soil nourishments that can be naturally discarded is an extraordinary method to diminish the measure of food you toss in the refuse. It is useful for the earth and can give you some helpful fertilizer for the nursery. Fertilizer will treat your dirt and help with any cultivating you do.

- Use food scraps, for example, products of the soil, espresso beans, eggshells, nutshells and tea sacks.
- Do not discard meat, dairy or oils along these lines.
- Add the food waste to cardboard, paper, vegetation and other natural materials in your fertilizer heap. Blend it in with soil and earth so the food can be separated.
- When you add new material to the heap, turn it with a nursery fork or other instrument to give new oxygen access to help with the treating the soil procedure.
- If you don't have a yard, you can even now compost at home with a night crawler ranch.

Visit Your Local Recycling Centre

On the off chance that you don't have the space, or are unsure about fertilizing the soil at home, you can in any case manage your food squander mindfully by visiting your nearby reusing focus. Many reusing focuses will have offices to manage food squander and fertilizing the soil. For the most part, you will have the option to bring your food squander and either leave it with somebody there, or dump it into the suitable compartment.

- Be sure you check the particular rules for your inside before you go.

- You may need to isolate your food squander with a specific goal in mind before you take in it.
- Make sure you comprehend what food squander they acknowledge and what they don't.
- For model, they won't take meat, yet they will take natural waste, for example, products of the soil.
- Your neighbourhood government will have the option to give subtleties on reusing offices close to you.

Use a Local Authority Food Waste Scheme

Contingent upon where you live, you might have the option to utilize a food squander plot that is controlled by your neighborhood authority. It is getting progressively basic for nearby specialists to give a little food squander fertilizing the soil canister to family units, which are then gotten close by the customary refuse assortment.

- If you are dubious, contact your nearby power to see whether your neighborhood is a piece of plan.
- Talk to neighbours to see whether they think about it, and how you can get included.
- In these plans, you will regularly be furnished with compostable packs to put your compostable food squander in.

2. Donating Unwanted Food

Determine What Food Goods are Suitable to Donate

On the off chance that you have food stopping up your cabinets that you won't eat, there is an option in contrast to simply throwing it in the garbage. Giving food to neighborhood noble cause, for example, food banks and soup kitchens, is an extraordinary method to ensure nothing goes to squander. In the event that you choose to do this, the main activity is figure out what sort of food is appropriate for you to give.

- Generally durable food, for example, tinned vegetables, soups, fish and meat are on the whole suitable.
- Low-sugar oat, tubs of nutty spread, raisins, and juice boxes are likewise exceptionally welcome.
- Avoid giving food in glass containers or holders. These may not be acknowledged as a result of the danger of them breaking.
- Remember you can likewise request that loved ones check whether they need anything as well.

Contact Local Charities

When you have a thought of what nourishments you have that might be reasonable to give, you have to look into some neighbourhood good cause. Quest for food banks and food drives in your general vicinity, and call up to request that how give. You can look for your neighbourhood food banks by utilizing the online hunt apparatus of national craving noble cause.

- There are extraordinary applications that help organizations to give undesirable food.
- As an individual, it is ideal to work with a set up good cause, either neighborhood or national.

Deliver it to the Food Centre

Cautiously get together your food and afterward take your bundles to your nearby food bank to convey them to the staff and volunteers there. They will be glad to see you and acknowledge your gift giving everything is pressed well and you have not included whatever isn't a suitable gift. While you're at the food drive you could discover somewhat more about the work they do. Regularly, they will be searching for new volunteers to help sort out and disseminate the gifts.

- If you have some extra time, why not engage with the foundation and volunteer with them.
- There is regularly a scope of chipping in circumstances accessible.

Donate Food from a Restaurant

You can also donate excess food to charity if you run a restaurant. Contact a local charity with details about the food you wish to donate, and they will send a truck or van to come and collect from the restaurant. These schemes may also enable you to donate perishable and prepared foods, which will then be frozen or delivered immediately to a local shelter.

Donate Food from a Grocery Store

You can likewise give abundance food from a supermarket or food wholesalers. The procedure is like that for eateries and lodgings. You should contact a nearby association and give them subtleties on the abundance food you might want to give. The association will at that point mastermind to get the food straightforwardly from you.

- You could turn into a corporate accomplice of a neighborhood or national food good cause in the event that you are an association that habitually has food squander.
- Being a corporate accomplice empowers you to make a standard game plan for gifts and may offer your business tax reductions.

3. Disposing of Non-Reusable Food Waste

Separate Spoiled Foods

You should act quick to manage any food that is ruined or that will turn sour rapidly. Such nourishments ought to be isolated from the remainder of your waste, kept in hard core plastic sacks and discarded rapidly. In the event that handy, put meats and different nourishments that decay rapidly into the rubbish on the day it will be gathered. Decaying food will pull in irritations and bugs.

- Tie meats and any crude nourishments you are discarding into plastic packs before you put them into your garbage sack. This will limit holes and smells.

- Ensure your garbage bin is completely made sure about and there is no smell that could draw in bugs.
- Dispose of your meat rapidly to keep away from any possible issues with slimy parasites.

Incinerate Items with Low Water Content, Such as Chicken Skin. (High-Water Content Items May Explode.)

- Use an indoor chimney or outside flame broil/fire pit for this.
- A wood oven can likewise be utilized, yet add the food to burn to the fuel region, not the cooking zone.
- Do NOT utilize a gas oven, as this will create an excessive amount of smoke inside.
- Try to do this when you would ordinarily assemble a fire in any case, to abstain from utilizing additional fuel. For instance, after an excursion, the pieces can be scorched on similar coals that were utilized to prepare the food (obviously making sure to soak everything in water before leaving the outdoors zone).
- Some cinders will remain, so discard those as you ordinarily do, when they cool.

Flush Items Down the Sink or Toilet

- For delicate things other than oils and fats, they might be slashed into little pieces and flushed down the sink channel, or bigger parts might be flushed down the latrine.
- Use this technique for delicate food scraps, similar to a spoiled tomato, versus unbending things, similar to bones.
- This elective is for the individuals who come up short on a waste disposal unit.

Collect Oils and Fats in a Container

Discard cooking oils and fats by gathering them in a container, tub, or other holder that you wouldn't fret discarding. Try not to pour hot oil or fat from cooking meats down your sink channel. This will mess plumbing up that might be costly. You ought to consistently discard oil and oil in the refuse, not down the channel.

- Throw the container of fat into the junk when it gets full. Try not to reuse the container.
- You can likewise utilize extra fat (or suet) to make fat balls for your nursery feathered creature feeder.
- Mix the fat with some dry kitchen scraps, for example, porridge oats, and leave it to set in the cooler short-term.
- Once it is hard you can balance it from a tree or spot it in a birdfeeder.

Use Your Garbage Disposal

On the off chance that you have a waste disposal appended to your sink, use it to discard food when you are tidying up after a supper. Scratch food squander into your channel and turn on the removal while running virus water. Tune in for the crushing of your food squander. Mood killer the removal and the water when you hear the cutting edges come back to their typical, void position.

- Remember not to place anything in your removal that isn't biodegradable.
- Don't put anything like glass, metal, plastic or paper in there.
- Don't pour oil or oil in your waste disposal.
- Don't put expandable food, similar to rice or pasta, in there.

Don't Put Food Waste into a Septic System

Putting food squander in your septic framework ought to be stayed away from. In the event that you have a septic framework take care not to wash food scraps, espresso beans, oils or fats down the sink. The more solids that make it into your take, the more frequently it should be siphoned.

- If you have a processor or waste disposal in your home attempt to constrain the sum you use it however much as could be expected.
- Using a waste disposal may influence the guarantee on your septic framework.

Know What Foods You can Just Throw Out

There are some foods that are inappropriate for composting or recycling, such as dried pasta, rice, bread and other grains. Dried foods like pasta and rice can be good to donate to food banks, and they generally have a very long shelf-life, so you shouldn't have to throw much away.

- If you do find some really old pasta or rice at the back of the cupboard you can just toss it in the trash.
- You may be tempted to feed stale bread to birds in your garden, but it has little nutritional value. Mouldy bread can even harm birds.
- Dairy products cannot be recycled or composted and should just be thrown in the trash.

1. Storing Food Scraps for Later Dispos

Be careful about putting out waste for disposal too early

It's not always possible to dispose of food immediately, such as if garbage collection is weekly. Leaving it in the garbage until then can generate smells, attract animals, and breed flies.

Use A "Slop Jar"

Ideally it should be made of glass, such as a pickle or spaghetti sauce jar, with a tight-fitting lid.

- Plastic jars can also be used, but may leak some odor.
- Sealable metal cans, such as coffee cans, can be used, but may create a rust ring if left in sink.
- Cardboard containers, like the cylinder's oatmeal comes in, should be avoided. They may leak when they get wet from decomposing food.
- Ideally the container ought to be little enough for a solitary use. Try not to resume it later, as this will let scent and conceivably flies or natural product flies out, except if it has been kept solidified.

Alternatively, Freeze Items For Later Disposal

This will stop the decay procedure and slaughter any creepy crawlies or hatchling. This strategy can either be utilized related to a slop container, or, more than likely huge things, similar to a watermelon skin, can be put in the cooler entirety. Be that as it may, it's anything but difficult to overlook these things on trash day, so a Post-it note on the garbage bin is a decent update.

5. Reducing Your Food Waste

Store Food Effectively

Find a way to assist you with lessening the measure of food squander you produce in the long haul. Regularly on the off chance that we store food mistakenly it turns sour before now is the right time, or it has an exceptionally restricted life expectancy. By setting aside the effort to store food all the more adequately you can squander less and show signs of improvement esteem for cash.

- Freeze any new nourishments that you won't utilize immediately so you can utilize them sometime in the not too distant future.
- Consider freezing extras, for example, soups, stews and pastas.
- Make sure the entirety of your stockpiling holders are secure and kept in the suitable climactic conditions.
- For model, a few nourishments should be kept in the refrigerator, and others ought to be in kept in a cool, dry spot.

Shop Smart

A basic method to diminish your food squander is to purchase less food. Focus on the amount you typically discard, and attempt to make a few acclimations to your basic food item shop as needs be. An extraordinary method to do this is to design out your dinners for the week and afterward simply purchase what you have to make those dishes.

- Be careful about uncommon offers and get one-get without one arrangement.
- If you can't store the additional food, inquire as to whether you will simply wind up discarding it.

Make the most of Leftovers

Another incredible method to eliminate food squander, and truly get the greater part of the food you purchase, is to benefit as much as possible from your extras. You can transform extras into additional dinners or tit bits, or use them to make stocks and stews. Search for plans that utilization what you have left. Expect to utilize all of food that you cooked. You can discover a large group of accommodating plans for extras on the web.

- Preserve or can any products of the soil that you have an excess of.
- Be sure you use extras securely. Pack them appropriately in water/air proof compartments, and keep them in the ice chest or cooler.
- Use extras inside two days, and don't warm food more than once.

Chapter 3

Food Costing

Baker's Formulas & Recipes

Plans and recipes are principal instruments of any kitchen and bakeshop. They show fixings to be bought and put away. They give estimating and readiness directions for the things to be delivered. What's more, they are the focal point of other administration apparatus and strategies, including changing amounts and deciding expenses.

Here we are acquainting you with bakeshop creation through a conversation of the sorts of estimations and scientific computations important for preparing and of the fundamental procedures regular to almost totally heated products.

Equations and Measurement

A formula is a lot of guidelines for creating a specific dish. So as to copy a merited planning, it is important to have an exact record of the fixings, their sums and the manner by which they are joined and cooked. This is the reason for formula.

Bakers for the most part talk about formulas as opposed to plans which seems like a science lab than a food creation in light of the fact that the bakeshop is a lot of like a science lab, both in the logical exactness of the systems and in the perplexing responses that happen during blending and preparing.

Note there are no careful principles about the utilization of the word recipe. A few dough punchers utilize the term for flour merchandise just while alluding to plans for things, for example, cake cream, natural product fillings and pastry mousses. Different bread cooks are prone to call all plans equations.

The essential capacity of an equation, obviously, is to give a lot of fixings and amounts for making an item. Yet, an equation is valuable for related purposes moreover. A composed recipe gives a method for changing amounts and yields and deciding expenses. These capacities require the utilization of math. Systems for working equation math are the fundamental focal point of this segment.

In spite of significance of composed equations and plans, they have numerous impediments, also. Regardless of how point by point an equation is, it expects you as of now have certain information – that you comprehend the wording it utilizes, for instance, do you realize how to quantify fixings.

Prior to experiencing explicitly about heating equations, we should quickly think about plans as a rule. Numerous individuals accept that figuring out how to cook implies just learning plans. An expert cook then again figures out how to function by acing a lot of fundamental methodology. A formula is a method of applying fundamental procedures to explicit fixings.

The fundamental reason for learning essential cooking standards isn't to have the option to cook without plans yet rather to comprehend the plans you use. Each formula expect that you have certain information, so you can comprehend the directions and tail them effectively. Your principle concern ought to learn systems that you can apply to any equation.

Bakers utilize a generally modest number of fundamental blending methods to get ready mixture and hitter. Hence, a dough puncher's recipe may comprise just of a rundown of fixings and amounts and the name of the blending strategy. A prepared bread cook can create a completed item with this data alone. Truth be told regularly the name of the blending technique isn't even essential in light of the fact that the pastry specialist can tell structure the fixings and their extents which blending strategy is required. To acclimate you to along these lines of working, and to underline the significance of learning fundamental blending strategies well, the vast majority of the equation in this book demonstrate the name of the necessary blending technique without rehashing the means for every recipe. For each situation, you should audit the fundamental methodology varying before utilizing equation.

A few plans gracefully almost no data and some flexibly extraordinary arrangement. Be that as it may, regardless of how nitty gritty it is, a composed formula can't disclose to you everything and some judgment by cooks should consistently make modifications for fixings item variety some carrot are better than others, a few shellfish are saltier than others, etc.

In the bakeshop, there is less item variety. For instance, flour, yeast, sugar, spread, and other fundamental fixings are really steady, particularly when bought from a similar source. In any case, may another factor can't be represented when composing a formula. For instance:

Equipment differs from bakeshop to bakeshop. For instance, various blenders just as various sizes of blenders process mixture distinctively and stoves change in their preparing properties.

It is difficult to give precise directions for some procedures. For instance, a bread recipe may demonstrate a blending time, however the specific time required for a specific group will differ. The bread cook must have the option to decide by the vibe and surface of the batter when it has grown appropriately.

Standardized Recipes and Formula

A normalized formula or recipe is a lot of directions portraying the manner in which a specific foundation readies a specific thing. As it were, it is a redone formula created by an activity for the utilization of its own cooks, baked good culinary specialists and pastry specialists, utilizing its own hardware, to be sold or served to its own benefactors.

Equation groups vary from activity to activity, however almost every one of them attempt to incorporate however much exact data as could reasonably be expected. The accompanying subtleties might be recorded:

- Name of formula
- Yield, including all out yield, number of segments and precise part size.
- Ingredients and accurate sums, recorded inorder of utilization.
- Equipment required, including estimating gear, dish sizes, parceling hardware, etc.
- Directions for setting up the dish. Bearings are kept as straightforward as could be expected under the circumstances.
- Preparation and cooking times.
- Directions for holding the item among readiness and administration.
- Directions for assigning, plating and embellishing.
- Directions for putting away extras.

As should be obvious, a portion of these focuses apply more to the baked good or treat station in a café than they do to retails bread shops. Bread plans don't require guidelines for plating and embellishing, for instance. All things considered the essential standards apply to pastry shops just as to eatery kitchens.

The Functions of Standardized Formulas

An activity's own plans are utilized to control creation. They do this in following manners:

They control quality. Standard equations are plans are itemized and explicit. This is to guarantee the item is the equivalent each time it is made and served regardless of who cooked it.

They control amount. To begin with, they show ideal amounts for each fixing and how to quantify that amount. Second, they demonstrate accurate yields and bit sizes and how to quantify and serve the segments.

The Limitation of Standardized Formulas

Normalized equations have indistinguishable issues from all plans the issues we talked about before with respect to variety in fixings, hardware and ambiguity of directions. These issues can be decreased by composing the formula cautiously, yet they can't be dispensed with. Regardless of whether an activity utilizes great normalized plans, another representative creation a dish just because typically requires some management to ensure the person in question deciphers the guidelines a similar route as the remainder of the staff. These confinements don't refute normalized plans. On the off chance that anything, they make careful bearings considerably more significant. In any case, they do imply that experience and information are still significant.

Instructional Recipe and Formulas

The equations are bit normalized however consistently recollect that a normalized recipe is hand crafted for a specific activity. The reason for a normalized equation is to direct and control the creation of a specific food thing. Bearings must be as finished and definite as could be expected under the circumstances.

The motivation behind the instructional equation is to acquaint you with essential preparing and cooking procedures. They give chances to you, to rehearse with explicit fixings, the overall systems you have learned.

On the off chance that you look at any of the recipe, you will see they don't contain all the highlights of normalized equations, as depicted in the past segment. Specifically, you will see the accompanying contrasts:

1. **Instructions for arrangement -** In many cases, equations follow a conversation of an essential strategy. The equations are instances of the overall system and they give you involvement with applying what you have realized. The data you are given in the recipe guidelines is expected basically to urge you to think and to get familiar with a strategy, not simply to turn out an item.
2. **Variation and discretionary fixings -** Many recipes are trailed by variety. There are in reality entire equations given in curtailed terms.

Giving recipes as variety as opposed to as independent equations urge us to see the examples behind every equation. Once more, this is a learning method and not simply equations. It's a method to grow additionally comprehension of what we are doing on the off chance that we see coconut cream pie and chocolate pudding as varieties of a similar essential strategies as opposed to as discrete disconnected equations.

Simply recall that may your teachers may have their own varieties, or they may wish to make changes in the fundamental recipes so as to tech your specific focuses. Not at all like normalized equations instructional recipes are not engraved in stone.

Understanding Formulas and Recipes

Prior to beginning creation, one should know or gain proficiency with the whole formula cautiously and totally. Culinary specialists consider these development arrangements their mise en place. A decent mise en place is fundamental for proficient activity of a bakeshop or kitchen. Coming up next are a portion of the errands make us to peruse the formula and prepare for creation.

Equation Modifications:

Determine the yield of the printed formula and choose if it needs change. On the off chance that you have to change over the formula to an alternate yield, do all the math in advance.

Determine if some other changes are required, for example, fixing replacements, to get the ideal outcome. Record them.

Fixings:

Assemble and measure all fixings. In the event that all fixings are scaled ahead of time, creation can go rapidly and without interference. Additionally, it's smarter to discover ahead of time in the event that you need more of a fixing so you can get more before beginning creation.

Prepare all fixings as essential, for example, filtering flour, isolating eggs and carrying margarine to room temperature. A considerable lot of these means are demonstrated in the formula however many may not be. Proficient recipes regularly expect the accomplished pastry specialist knows, for instance, that margarine ought to be expelled from the fridge ahead of time so it is sufficiently delicate to be utilized in creaming technique cake players.

Strategies:

- Read the whole strategy or technique for readiness cautiously and ensure you get it.
- If a blending strategy is demonstrating just by name, for example, Creaming Method, gaze upward and audit the methodology on the off chance that you have to revive your memory. Ensure you see each progression of the overall strategy and how to apply it to your particular equation.
- Look up any terms or catchphrases you don't have a clue.

Devices and Equipment:

Determine what hardware you need. Required gear is commonly recorded in normalized plans however not in plans from other source. Peruse each progression of the technique and compose which instruments and hardware you need in each progression.

- Assemble all devices and hardware.
- Prepare hardware varying. For instance, line sheet container with material, oil cake skillet, preheat broilers.

Measurement

One of the essential elements of a recipe is to show the fixings and right amounts or estimations to be utilized to make an item.

Fixings are quite often said something the bakeshop as opposed to estimated by volume since estimation by loads more precise. Precision of estimation, as we have said is fundamental in the bakeshop. In contrast to home preparing plans, an expert dough puncher's equation won't call for 6 cups flours, for instance.

To exhibit to yourself the significance of weighing as opposed to estimating by volume, measure 1 cup flour in two different ways

a) filter some flour and gently spoon it into a dry measure, level the top and gauge the flour.

b) scoop same unshifted flour into a similar measure and pack it daintily. Level the top and weight the flour. Note the distinction. No big surprise home plans which as a rule show volume proportion of dry fixings taxi be so conflicting.

The baker's term for weighing ingredients is scaling.

The following ingredients and only these ingredients may sometimes be measured by volume, at the ratio of 1 Litre=1000 miligram

- Water
- Milk
- Eggs

Volume measure is frequently utilized when scaling water for little or medium estimated groups of bread. Results are commonly acceptable. Be that as it may, at whatever point exactness is basic, it is smarter to gauge. This is on the grounds that 1 liter water really gauges same 1000 grams (this figure changes with the temperature of the water).

For comfort, volume proportions of fluids are much of the time utilized when items other than heated flour products, for example, sauces, syrups, puddings and custards-are being made.

Unit of Measure

Food cost estimation is a significant factor of Food administration business. Cooks ought to know about the nuts and bolts of food cost and its count, regardless of whether they are not answerable for the bookkeeping or account sides. The vast majority of costs will fall into two classes variable (immediate or roundabout) and fixed.

Variable Cost – incorporated those costs that change with the quantity of units made, for instance crude fixings, preparing costs, pressing cost, work cost, and so on.

Fixed expense – incorporated those costs that don't change every once in a while and the cost happen same notwithstanding number of units made, for instance kitchen equipment's, utilities, the executives pay, and so forth.

The arrangement of estimation utilized is muddled. Experience underneath table for reference:

METRIC and U.S. MEASUREMENT EQUIVALENCIES

METRIC SYSTEM OF MEASUREMENT:

Basic unit of LENGTH........................METER (m)..................100cm = 1000 mm = .001 km
Basic unit of MASS (weight)........................GRAM (g)..................001 kg = 1000 mg
Basic unit of CAPACITY (liquid measurement)..................LITER (L)..................1000 ml = .001 kl

Important Metric System Prefixes:

kilo	=	1,000	deci	=	0.1	tera	=	1,000,000,000,000
hector	=	100	centi	=	0.01	giga	=	1,000,000,000
deca	=	10	milli	=	0.001	mega	=	1,000,000

LENGTH			WEIGHT			CAPACITY		
1 meter (m)	=	100 centimeters (cm)	1 gram (g)	=	1000 milligrams (mg)	1 L	=	1000 mL
1 cm	=	0.01 m	1 kilogram (kg)	=	1000 grams	1 mL	=	0.001 L
1 m	=	1000 millimeters (mm)	1 mg	=	0.001 g	1 k	=	1000 L
1 mm	=	0.001 m	1 g	=	0.001 kg			
1 cm	=	10 mm						
1 mm	=	0.1 cm						
1 kilometer	=	1000 m						
1 m	=	0.001 km						
1 megameter	=	1,000,000 m						
1 hectometer	=	100 m						
1 dekameter	=	10 m						
1 decimeter	=	1/10 m						
1 cm	=	1/100 m						
1 mm	=	1/1000 m						
1 micrometer	=	1/1,000,000 m						

U.S. CUSTOMARY SYSTEM OF MEASUREMENT:

LENGTH			WEIGHT			CAPACITY			AREA		
1 foot	=	12 inches	1 lb	=	16 oz	1 cup	=	8 fl oz	1 ft^2	=	144 in^2
1 yard	=	3 feet	1 ton	=	2,000 lbs	1 pint	=	2 cups	1 yard^2	=	9 ft^2
1 yard	=	36 inches				1 quart	=	4 c/2 pts	1 acre	=	43,560 ft^2
1 mile	=	5,280 feet				1 gal	=	4 quarts	1 mile^2	=	640 acres

APOTHECARIES' WEIGHTS and MEASURES:

1 dram	=	60 grains (gr)	1 fluid dram	=	60 minims
1 ounce	=	8 drams	1 ounce	=	8 fluid drams
1 pound (lb)	=	12 ounces	1 pint (pt)	=	16 oz
1 ounce	=	480 gr	1 quart	=	2 pints
1 fluid oz	=	480 minims	1 gallon	=	4 quarts
1minim	=	1 gr			

Basic Units

There is one essential unit for each sort of estimation.

- The gram is the essential unit of weight.
- The liter is the essential unit of volume.
- The meter is the essential unit of length.
- The degree Celsius is the essential unit of temperature.

Bigger or littler units are made basically by increasing or isolating by 10, 100, 1000, etc. These divisions are communicated by prefixes. The one you have to know are:

Kilo = 1000

Deci = 1/10 or 0.1

Centi = 1/100 or 0.01

Mili = 1/1000 or 0.001

A kilogram is somewhat multiple pounds (approx. 2.2 pounds)

A gram is around 1/30 ounce. A half-teaspoon of flour gauges somewhat less than 1 gram

A liter is marginally more than 1 quart

A decilitre is marginally not exactly ½ cup

A centiliter is around 2 teaspoons

A meter is marginally longer than 3 feet

A centimeter is around 3/8 inch

0 degree Celsius is 32 Fahrenheit (the point of solidification)

100 degree Celsius is 212 Fahrenheit (breaking point)

Once you know the basic units, you don't need complicated tables.

Metric Formulas and Recipes

American industry will most likely receive the decimal measuring standard sometime in the future. Numerous formula authors are getting a head start by including metric counterparts. Accordingly, you will see plans calling for,

for instance: 500 g flour, 150 g spread, or a preparing

Metric Units

Length

1 centimeter = 10 millimeters
1 meter = 100 centimeters
1 meter = 1,000 millimeters
1 kilometer = 1,000 meters

Weight

1 gram = 1,000 milligrams
1 kilogram = 1,000 grams

Capacity

1 liter = 1,000 milliliters
1 kiloliter = 1,000 liters

temperature of 191°C. No big surprise a few people fear the decimal standard for measuring!

Kitchens in nations that utilization the decimal measuring standard don't work with such unreasonable numbers anything else than we regularly use figures like 1 lb 1/4 oz flour, 2.19 oz spread, or a preparing temperature of 348°F. That would invalidate the point of the decimal standard for measuring, which is to be straightforward and handy. In the event that you get an opportunity to take a gander at a French cookbook, you will see pleasant, round numbers such as 1 kg, 200 g, and 4 dL.

To give careful counterparts would require utilizing abnormal, unfeasible numbers. In the event that you have metric gear, utilize the metric units; on the off chance that you have U.S. hardware, utilize the U.S. units. As noted, before, once in a while should you need to stress over changing over between the two.

Generally, the complete yield of the metric recipes in this book is near the yield of the U.S. equations, while keeping the fixing extents the equivalent. Lamentably, it isn't generally conceivable to keep the extents precisely the equivalent in light of the fact that the U.S. framework isn't decimal-based like the decimal measuring standard. Sometimes, the metric amounts produce marginally various outcomes because of the shifting extents, yet these distinctions are typically incredibly little.

Measuring by Weight

A decent parity scale ought to be precise to 5 g or 1/4 oz (0.25 oz). Dry fixings weighing under 5g can be scaled by genuinely isolating bigger amounts into equivalent bits. For instance, to scale 2.5g , first weigh out 10g, at that point isolate this into four equivalent heaps utilizing a little blade.

For fine baked good work, a little hitter y-worked computerized scale is frequently more helpful than a huge parity scale. A decent advanced scale is generally cheap. It can quickly gauge amounts to the closest 2 g. Significantly more delicate scales are accessible at a to some degree more significant expense. Most advanced scales have a zero, or tare, button that sets the demonstrated load to zero. For instance, you may set a compartment on the scale, set the weight to zero, include the ideal amount of the main fixing, again set the weight to zero, include the subsequent fixing, etc. This speed the weighing of dry fixings to be filtered together, for instance. Be cautious, be that as it may, when utilizing this technique, instead of gauging fixings each in turn. On the off chance that you include a lot of one fixing, you will probably need to dispose of the entire blend and start once more.

At the point when little amounts of things, for example, flavors are required in recipes, a rough volume proportionate (generally in parts of a teaspoon) is likewise included. In any case, recall that cautious burdening a decent scale is more exact.

AP Weight and EP Weight

In the hot kitchen, cooks are routinely worried about the cutting yield of vegetables, natural products, meats, and different fixings. For instance, 1 kg apple yield substantially less than 1 kg cut, stripped apple. In the bakeshop, pastry specialists need not be worried about cutting yield of the fixings they utilize most: flour, sugar, fats, etc. In any case, it is essential to have the option to make the best possible yield figurings when working with new natural products. What number of kilograms of entire apples should the dough puncher request, for instance, if 5 kg stripped, cut apples are required?

The rate yield of a natural product or vegetable shows, by and large, the amount of the AP weight (as bought weight) is left in the wake of cutting to create the prepared to-cook thing, or EP weight (eatable part weight).

Procedure for Using a Baker's Balance Scale

The principle of using a baker's scale is simple. The scale must be balanced before setting the weights and it must balance again after scaling. You can use digital weighing scale where you can just place ingredients on top of scale to measure and check on scale display to get accurate weight.

Trimming yield of ingredients, they use most-flour, sugar, fats and so on. However, it is important to be able to make the proper yield calculations when working with fresh fruits. How many Kg of whole mangoes must the baker order, for example, if 5 kg peeled, sliced mangoes needed?

The percentage yield of a fruit or vegetable indicates, on the average, how much of AP weight (as purchased weight) is left after trimming to produce the ready-to-cook item or EP weight (edible portion weight).To determine the percentage yield of a fruit, follow below steps:

1. Weigh the thing before cutting. This is the AP weight.
2. Trim and strip the thing as important to get the eatable part.
3. Weight the cut thing. This is EP weight.
4. Divide the EP weight by the AP weight.
5. Multiply this number by 100 to get the percentage

The most accurate yield percentages are the ones you calculate yourself, because they are based on the items you actually use in your bakeshop. For approximate or average yield percentages of most commonly used fruits.

Once you have a yield percentage for an item, save this number to refer to as necessary. You can use this figure to do two basic calculations.

1. Calculating yield – example, you have 10 kg AP apple. Yield after trimming is 75%. What will your EP weight be? First, change the percentage to a decimal number by moving the decimal point two places to the left. 75% = 0.75 Multiply the decimal by your AP weight to get EP yield – 10 Kg X 0.75 = 7.5 Kg
2. Calculating amount needed – example, you need 10 kg EP apple slices. What amount of untrimmed fruit do you need?

Change the percentage to a decimal number 75% = 0.75

Divide the EP weight needed by this number to get the AP weight

10/0.75= 13.33 kg

Baker's Percentages

Formula: -

$$\frac{\text{total weight of ingredient}}{\text{total weight of flour}} \times 100\% = \%\text{ of ingredient}$$

The most significant data passed on by a baker's recipe is the proportion of the fixings to one another. For instance, in the event that you know a specific bread dough requires precisely 66% as much water as possible, consistently know the specific measure of water to add to the flour, regardless of whether you are making an enormous or a little amount. Proportions are the least difficult and most fundamental method of communicating an equation.

Baker utilize a basic however flexible arrangement of rates for communicating their equations. Pastry specialist's rates demonstrate the measure of every fixings utilized as a level of the measure of flour utilized.

To put it in an unexpected way, the level of every fixing is its all out weight separated by the heaviness of the flour increased by 100% or absolute load of fixing/all out weight flour X 100% = % of fixing

In this way flour is consistently 100%. On the off chance that two sorts of flour are utilized, their absolute is 100%. Any fixing that gauges equivalent to the measure of flour utilized is likewise given as 100%.

It would be ideal if you recall these numbers don't allude to the level of the complete yield of these rate numbers will consistently be more prominent than 100%.

Baker's rates make it simple to see initially the fixing proportions and, consequently the essential structure and synthesis of the mixture or player. What's more, they make it simple to adjust the equation for any yield, as you will find in a later segment. A third favourable position is that solitary fixings might be changed and different fixings included without charging the entire detailing. For instance, you add raisins to a biscuit blend recipe while keeping the level of the various fixings the equivalent.

Dough puncher's rates are the most essential method of communicating a recipe, so they are additionally a helpful instrument for growing new equations. When conceiving another recipe, a cook contemplates the best apportion of fixings, as showed by rates. When the correct proportions are built up, the dough puncher would then be able to make an interpretation of them into loads, with the goal that the equation can be tried. The vast majority of the recipes in this book were formulated along these lines.

Plainly, a rate framework dependent on the heaviness of flour can be utilized just when flour is a significant fixing, as in breads, cakes and treats. Be that as it may, the guideline can be utilized in different recipes too by choosing a significant fixing and setting up 100%. Numerous bread cooks utilize the rate framework for flour merchandise just (mixtures and players) yet it is useful to expand the advantages of this framework to different items. In this book, at whatever point a fixing other than flour is utilized as the base of 100%, this is demonstrated at the head of the recipe over the rate section. For instance, the equations for almond filling. These plans designate "almond glue at 100%" and the heaviness of the sugar, eggs and different fixings are communicated as rates of the heaviness of the almond paste.

Formula Yields

Yields for the recipes in this book are demonstrated in one of two different ways. Much of the time, the yields are given as an aggregate of the fixing's amounts. For instance, test recipes on next page, the yield discloses to us how much cake hitter the equation makes. This is the figure we have to know to scale the hitter into dish. The genuine load of the heated cake will differ contingent upon container size and shape, stove temperature, etc.

Different recipes of this sort, in which the yield is the all out weight of the fixings, incorporate equations for bread mixtures, espresso cake filings, cake batters and treat batters.

In certain equations, be that as it may, the yield isn't equivalent to the absolute load of fixings. For instance, when sugar and water are bubbled to make a syrup, about a large portion of the water vanishes. In this manner, the real yield is not exactly the all out weight of the fixings.

At the point when yield isn't equivalent to the all out weight of the fixings, the yield is showing over the fixings list as opposed to beneath it.

Likewise, it would be ideal if you not that all yields, including rate aggregates, are adjusted to the following lower entire number. This dispenses with inconsequential parts and makes perusing simpler. Basic Formula and Recipe Conversion

Except if you are working in an activity that utilizes just its own normalized recipes, you every now and again will be required to change over equations to various amounts. For instance, you may have an equation for 10 kg batter yet need just 4 Kg.

Changing over equations and plans is a significant strategy. There is no 'best" respect compose plans for, as each activity, each school and each individual have various necessities.

About everybody naturally can twofold an equation or cut it down the middle. It appears to be more convoluted, though, to change a recipe from 10 t0 18 kg, state, or from 20 to 12 qt. As a matter of fact, the rule is the very same: You increase every fixing by a number called a transformation factor, as in the principal technique given here.

The two strategies are general ones. These techniques are likewise utilized for plans in the hot kitchen. The second of the two, for changing segment sizes, is pertinent to a portion of the sweet plans in this book however is, obviously, not utilized when segment sizes are not given, as in recipes for bread batter. At that point, in the following area, we take a gander at strategies for making figuring's with preparing rates.

Technique for Calculating the Weight of an Ingredient when the Weight of Flour is Known

1. Change the fixing rate to decimal structure by moving the decimal direct two spots toward the left.
2. Multiply the heaviness of the flour by this decimal figure to get the heaviness of the fixing.

Example:

A formula calls for 20% sugar and you are using 10 kg flour. How much sugar do you need?

20% = 0.20

10 kg X 0.20 = 2 kg sugar

Procedure for Converting a Formula to a New Yield

1. Change the all out rate to decimal structure by moving the decimal guide two spots toward the left.
2. Divide the ideal yield by this decimal figure to get the heaviness of flour required.
3. If essential, adjust this number to the following most elevated figure. This will consider looses in blending, cosmetics and panning and it will make figuring simpler.
4. the heaviness of flour and remaining fixing rates to ascertain the loads of different fixings, as in the past technique.

Ingredients	Weight	%
Flour	2500g	100%
Sugar	2500g	100%
Baking Powder	125g	5%
Salt	63g	2.5%
Shortening	1250g	50%
Milk	1500g	60%
Egg Whites	1500g	60%
Total	9438g	377.5%

Example: In the sample cake formula in the table, how much flour is needed if you require 3kg cake batter?

377.5% = 3.775

3000 gram/3.775 = 794.7 gram, rounded off to 800 grams.

Problems in Converting Formulas

Generally changing over preparing equations to various yields functions admirably. For whatever length of time that fixing proportion remain the equivalent, you are making a similar mixture or hitter. In any case, when you make exceptionally enormous transformations state from 2 kg mixture to 100 kg-you may experience issues all in all the major is in one of the accompanying classes.

Surface and Volume

In the event that you have considered calculation, you may recollect that a solid shape with a volume of 1 cubic foot has a top surface territory of 1 square foot. Yet, on the off chance that you twofold the volume of the 3D square, the top surface zone isn't multiplied yet is in actuality just about 1% times as huge.

What on the planet, you ask does this have to do with cooking? Think about the accompanying model:

- Assume you have a decent formula for ½ liter desert sauce, which you regularly make in a little pot. You need to make 16 liter sauce, so you increase all fixings by a change factor of 32 and make the sauce in a steam pot. Amazingly in addition to the fact that you end up with more sauce than you expected, yet it turns out rather flimsy and watery. What was the deal?
- Your changed over formula has 32-fold the amount of volume to begin, however the measure of surface zone has not expanded close to so a lot. Since the proportion of surface territory to volume is less, vanishing is less. This implies less decrease and less thickening happen and the flavours are not as concentrated. To address this issue, you would need to either stew the sauce longer or utilize less fluid.
- The surface/volume issue appears in the distinction between making a solitary portion of bread at home and making a huge amount of bread in the pastry shop. The home bread pastry specialist utilizes warm water to make a bread mixture and must figure out how to keep the batter sufficiently warm so it ages appropriately. The apportion of surface zone to volume in a modest quantity of batter is so high the mixture cools rapidly. The business pastry specialist, on the opposite regularly utilizes ice water when making bread batter to guarantee the mixture doesn't turn out to be excessively warm. The proportion of surface to volume is low and the mixture holds the warmth created by blending.
- When making enormous modifications in equation yields, you should likewise decide whether changes in strategy or fixings rate are required.

Equipment

At the point when you change the size of a recipe, you should frequently change the gear, as well. This change frequently implies the formula doesn't work similarly Bakers and cooks must have the option to utilize their judgment to envision these issues and to alter their strategies to keep away from them. The model simply given, of cooking a huge group of treat sauce in a steam pot, is among the sorts of issues that can emerge when you change cooking utensil.

Different issues can emerge due to blenders or other preparing gear. For instance, in the event that you separate a batter equation to make just a little amount, you may discover there is so little mixture in the blending machine that the mixers don't mix the fixings appropriately.

Or on the other hand you may have a formula for a biscuit player you for the most part make in little amounts, blending the hitter by hand. In this way, you utilize a blender however keep the blending time the equivalent. Since the blender carries out the responsibility so productively, you overmix the hitter and end up with low quality biscuits.

Numerous mixings and blending employment should be possible just be hand. This is simple with little amounts however troublesome with huge groups. The outcome is regularly an inside item. Then again, some high-quality items are better in the event that they are created in enormous bunches. It is hard, for instance, to make a little cluster of puff baked good since batter can't be rolled and collapsed appropriately.

Selection of Ingredients

Notwithstanding estimating, there is another essential principle of precision in the bakeshop; utilize the specific fixings indicated. You should be cautious while choosing fixings as pastry specialist's equations are adjusted for explicit fixings and don't work. Substitute things probably won't give same quality or taste. For instance, there are various flours or shortenings and bread flour can't supplant with cake flour or customary shortening with emulsified shortening. They won't work a similar way.

Periodically, a replacement might be made, for example, dynamic dry yeast for compacted yeast, yet not without changing the amounts and rebalancing the recipe.

Cost Calculations

Food administration activities are organizations. Gourmet experts and cooks must know about the nuts and bolts of food cost figurings, regardless of whether they aren't answerable for the administration of spending plans, solicitations and costs. This area talks about the most essential figuring.

Fixings Unit Cost

The straightforward food cost will be cost by unit; for instance, from complete expense to unit cost – Total Cost partition by Number of units = Unit cost

Details : - Often the purveyor's receipt demonstrates unit cost; for instance, 10 Kg mangoes at Rs. 30.00 per Kg, totalling Rs.300.00 (10 Kg X Rs.30.00). In different cases, you should make this computation, utilizing the accompanying equation:

All out Cost separate by Number of Units = Unit Cost

Example - 10 Kg mangoes cost of Rs. 300.00 per Kg, what will be the cost of per KG?

Rs. 300 partitions by 10Kg = Rs. 30.00 per Kg

EP Unit Cost

AP (As Purchase) Cost is all out cost paid for fixings and EP (Edible Portion) Cost is remembered yield for head of AP cost. For instance, you are paying for the whole organic products however you dispose of strips, seeds and pits.

Figuring of AP and EP amounts are vital not just for deciding amounts required for planning equations and plans yet in addition for deciding expenses. All things considered, when you purchase new natural product by weight, for instance, you are paying for whole organic product, regardless of whether you dispose of strips, centres and pits.

In above model, we confirmed that we are paying Rs.30.00 per AP Kg of mangoes. Be that as it may, we dispose of the strip and pit, so the expense per EP Kg is more noteworthy than Rs.30.00. we utilize the accompanying equation to figure the yield cost or EP unit cost: Formula Cost

To determine the cost of preparing a formula or recipe, we first determine the cost of each ingredient. Then we add the cost of all the ingredients to get the total cost of the formula.

When we have calculated the total cost, we can then determine the unit cost of the finished product. Unit may be any measure we require; per gram, per kilogram or per serving portion (portion cost).

For the most accurate costing, you should determine the number of units actually sold, rather than the unit yield of the formula. Keep in mind that ingredients or product lost through spillage or other waste must still be accounted for and paid for. Using units sold or served accounts for these costs.

The following general procedure explains the basic steps in calculating formula cost.

Procedure for Calculating Formula Costs:

1. List all ingredients and quantities of the formula as prepared.
2. Determine the EP unit cost of each ingredients
3. Convert the quantities in the formula to the same unit used for EP cost.
4. Calculate the total cost of each ingredient by multiplying the EP unit cost by the number of units needed. Round up the fractions of a paise to the next highest paise.
5. Add the ingredient costs to get the total formula cost.
6. To get the unit cost, divide the total formula cost by the number of units produced.

Example - Costing a Formula of biscuit dough

Ingredients	Amount	Amount in Converted Units	EP Unit Cost	Total
Flour	1/2kg 70gm	570gm	Rs. 0.04 per gram	Rs. 22.80
Whole Wheat Flour	1/2kg 70gm	570gm	Rs. 0.036 per gram	Rs. 20.52
Salt	22g	22g	Rs. 0.01 per gram	Rs. 0.22
Sugar	57g	57g	Rs.0.045 per gram	Rs. 2.57
Baking Powder	71g	71g	Rs. 0.30 per gram	Rs.21.30
Butter	397g	397g	Rs.0.47 per gram	Rs. 186.59
Milk	½ Litre 237 ml	737 ml	Rs.0.05 per ml	Rs 36.85
Total	2 kg 424g	2424g		Rs 290.85
Cost Per Gram				Rs. 0.12

Now weigh the biscuit in grams and get cost by multiplying above per gram cost to get the cost per unit.

Metric Conversion Factors

Temperature

To convert Fahrenheit to Celsius: Subtract 32 then multiply by 5⁄9.

Example: Convert 140°F to Celsius.

140 – 32 = 108

108 × 5⁄9 = 60°C

To convert Celsius to Fahrenheit: Multiply by 9⁄5 then add 32.

Example: Convert 150°C to Fahrenheit.

150 × 9⁄5 = 270

270 + 32 = 302°F

Weight

1 ounce equals 28.35 grams

1 gram equals 0.035 ounce

1-pound equals 454 grams

1 kilogram equals 2.2 pounds

Volume

1 fluid ounce equals 29.57 millilitres

1 millilitre equals 0.034 fluid ounce

Length

1-inch equals 25.4 millimetres

1 centimetre equals 0.39 inch

1-meter equals 39.4 inches

US Cups to Grams Conversions For Common Ingredients

Almonds (Flaked)

US cups	Amount in Grams
1/8 cup	10g
1/4 cup	20g
1/3 cup	20g
1/2 cup	45g
2/3 cup	55g
3/4 cup	65g
1 cup	85g

Almonds (whole)

US cups	Amount in Grams
1/8 cup	20g
1/4 cup	40g
1/3 cup	50g
1/2 cup	75g
3/4 cup	115g
1 cup	150g

Blueberries

US cups	Amount in Grams
1/8 cup	15g
1/4 cup	25g
1/3 cup	35g
1/2 cup	50g
2/3 cup	65g
3/4 cup	75g
1 cup	100g

Brown Sugar (Light)

US cups	Amount in Grams
1/8 cup	25g
1/4 cup	50g
1/3 cup	65g
1/2 cup	100g
2/3 cup	135g
3/4 cup	150g
1 cup	200g

Butter & margarine

US cups	Amount in Grams	
1/8 cup	28g	
1/4 cup	57g	
1/3 cup	75g	
1/2 cup	113g	
2/3 cup	150g	
3/4 cup	170g	6.25
1 cup	225g	

Buttermilk

US cups	Grams
1/8 cup	30g
1/4 cup	55g
1/3 cup	75g
1/2 cup	115g
2/3 cup	150g
3/4 cup	170g
1 cup	225g

Caster Sugar – Superfine Sugar

US cups	Amount in Grams
1/8 cup	30g
1/4 cup	55g
1/3 cup	75g
1/2 cup	115g
2/3 cup	150g
3/4 cup	170g
1 cup	225g

Chocolate Chips

Please note: As there are many different brands of chocolate chips that come in a variety of shapes and sizes, these are only approximate weights.

US cups	Amount in Grams
1/8 cup	19g
1/4 cup	37.5g
1/3 cup	50g
1/2 cup	75g
2/3 cup	100g
3/4 cup	113g
1 cup	150g

Cocoa Powder

US cups	Grams
1/8 cup	15g
1/4 cup	25g
1/3 cup	35g
1/2 cup	50g
2/3 cup	65g
3/4 cup	75g
1 cup	100g

Coconut – Shredded & Desiccated

US cups	Amount in Grams
1/8 cup	12.5g
1/4 cup	25g
1/3 cup	33g
1/2 cup	50g
2/3 cup	66g
3/4 cup	75g
1 cup	100g 3

Confectioners' Sugar – Powdered Sugar – Icing Sugar

US cups	Amount in Grams
1/8 cup	15g
1/4 cup	25g
1/3 cup	35g
1/2 cup	50g
2/3 cup	65g
3/4 cup	75g
1 cup	120g

Corn Starch – Cornflour

US cups	Amount in Grams
1/8 cup	15g
1/4 cup	25g
1/3 cup	35g
1/2 cup	50g
2/3 cup	65g
3/4 cup	75g
1 cup	100g

Cornmeal

US cups	Amount in Grams
1/8 cup	20g
1/4 cup	40g
1/3 cup	50g
1/2 cup	75g
3/4 cup	115g
1 cup	150g

Corn Syrup

US cups	Amount in Grams
1/8 cup	45g
1/4 cup	85g
1/3 cup	110g
1/2 cup	170g
2/3 cup	225g
3/4 cup	250g
1 cup	340g

Cream Cheese/Soft Cheese

US cups	Amount in Grams
1/8 cup	15g
1/4 cup	30g
1/3 cup	40g
1/2 cup	60g
2/3 cup	80g
3/4 cup	90g
1 cup	120g

Dark Brown Sugar – Muscovado Sugar

US cups	Amount in Grams
1/8 cup	25g
1/4 cup	50g
1/3 cup	65g
1/2 cup	100g
2/3 cup	135g
3/4 cup	150g
1 cup	200g

Dates (whole, Pitted)

US cups	Amount in Grams
1/8 cup	30g
1/4 cup	55g
1/3 cup	75g
1/2 cup	115g
2/3 cup	150g
3/4 cup	170g
1 cup	225g

Flour

US cups	Amount in Grams
1/8 cup	17g
1/4 cup	35g
1/3 cup	46g
1/2 cup	70g
2/3 cup	95g
3/4 cup	105g
1 cup	140g

Golden Syrup

US cups	Amount in Grams
1/8 cup	45g
1/4 cup	85g
1/3 cup	110g
1/2 cup	170g
2/3 cup	225g
3/4 cup	250g
1 cup	340g

Granulated Sugar

US cups	Amount in Grams
1/8 cup	15g
1/4 cup	25g
1/3 cup	35g
1/2 cup	50g
2/3 cup	65g
3/4 cup	75g
1 cup	200g

Honey

US cups	Amount in Grams
1/8 cup	45g
1/4 cup	85g
1/3 cup	110g
1/2 cup	170g
2/3 cup	225g
3/4 cup	250g
1 cup	340g

Jam – Jelly – Preserves

US cups	Amount in Grams	
1/8 cup	40g	1.4 oz
1/4 cup	80g	3 oz
1/3 cup	110g	4 oz
1/2 cup	165g	
2/3 cup	215g	
3/4 cup	245g	
1 cup	325g	

Lentils (Dry)

US cups	Amount in Grams
1/8 cup	25g
1/4 cup	50g
1/3 cup	65g
1/2 cup	100g
2/3 cup	135g
3/4 cup	150g
1 cup	200g

Maple Syrup

Please note: Since these ingredients can also be considered liquids, please see the volume conversion chart for liquid conversions.

US cups	Amount in Grams
1/8 cup	45g
1/4 cup	85g
1/3 cup	110g
1/2 cup	170g
2/3 cup	225g
3/4 cup	250g
1 cup	340g

Mincemeat (fruit)

US cups	Amount in Grams
1/8 cup	30g
1/4 cup	55g
1/3 cup	75g
1/2 cup	115g
2/3 cup	150g
3/4 cup	170g
1 cup	225g

Molasses

US cups	Amount in Grams
1/8 cup	40g
1/4 cup	80g
1/3 cup	110g
1/2 cup	165g
2/3 cup	215g
3/4 cup	245g
1 cup	325g

Oats (old fashion oats & porridge oats)

US cups	Amount in Grams
1/8 cup	17.5g
1/4 cup	35g
1/3 cup	46g
1/2 cup	70g
3/4 cup	105g
1 cup	140g

Parmesan Cheese (Grated)

US cups	Amount in Grams
1/8 cup	12.5g
1/4 cup	25g
1/3 cup	33g
1/2 cup	50g
3/4 cup	75g
1 cup	100g

Pasta (Dry)

US cups	Amount in Grams
1/8 cup	15g
1/4 cup	30g
1/3 cup	40g
1/2 cup	60g
2/3 cup	80g
3/4 cup	90g
1 cup	120g

Peanuts, Chopped

US cups	Amount in Grams
1/8 cup	15g
1/4 cup	30g
1/3 cup	40g
1/2 cup	60g
2/3 cup	80g
3/4 cup	90g
1 cup	120g

Peanuts (whole, Shelled)

US cups	Amount in Grams
1/8 cup	20g
1/4 cup	40g
1/3 cup	50g
1/2 cup	75g
3/4 cup	115g
1 cup	150g

Peanut Butter

US cups	Amount in Grams
1/8 cup	30g
1/4 cup	65g
1/3 cup	85g
1/2 cup	125g
2/3 cup	165g
3/4 cup	190g
1 cup	250g

Pecans (Chopped & Halved)

US cups	Amount in Grams
1/8 cup	15g
1/4 cup	30g
1/3 cup	40g
1/2 cup	60g
2/3 cup	80g
3/4 cup	90g
1 cup	120g

Pearl Barley

US cups	Amount in Grams
1/8 cup	25g
1/4 cup	50g
1/3 cup	65g
1/2 cup	100g
2/3 cup	135g
3/4 cup	150g
1 cup	200g

Raisins – Golden Raisins – Sultanas

US cups	Amount in Grams
1/8 cup	20g
1/4 cup	40g
1/3 cup	50g
1/2 cup	75g
3/4 cup	115g
1 cup	150g

Raspberries

US cups	Amount in Grams
1/8 cup	15g
1/4 cup	30g
1/3 cup	40g
1/2 cup	60g
2/3 cup	80g
3/4 cup	90g
1 cup	120g

Rice

US cups	Amount in Grams
1/8 cup	25g
1/4 cup	50g
1/3 cup	66g
1/2 cup	100g
2/3 cup	133g
3/4 cup	150g
1 cup	200g

Strawberries

US cups	Amount in Grams
1/8 cup	25g
1/4 cup	50g
1/3 cup	66g
1/2 cup	100g
2/3 cup	133g
3/4 cup	150g
1 cup	200g

Walnuts (chopped)

US cups	Amount in Grams
1/8 cup	15g
1/4 cup	30g
1/3 cup	40g
1/2 cup	60g
2/3 cup	80g
3/4 cup	90g
1 cup	120g

Primary Sidebar

1 cup equals 237 millilitres

1 quart equals 946 millilitres

1 litre equals 33.8 fluid ounces

Chapter 4

Bakery

BAKING AND PASTRY EQUIPMENT

Equipment Sanitation and Safety

Safety

Baking equipment can be dangerous.

Much of the equipment found in the bakeshop can inflict serious injuries if not used carefully and properly

Guidelines to use equipment:

- Never use a piece of equipment until you are thoroughly familiar with its operation and all its features.
- You must also learn to recognize when a machine is not operating correctly so you can shut it down immediately and report the malfunction to a supervisor.
- Be aware that not all models are alike. Each manufacturer introduces slight variations on the basic equipment.
- All deck ovens or all vertical mixers, operates on the same basic principles, each model is slightly different in the location of the switches.
- It is important to study the operating manual supplied with each item, or be taught by someone who already knows the item well.

Sanitation

- Thorough, regular cleaning of all equipment is essential.
- Most large equipment can be partially disassembled for cleaning.
- Read the operating manual, which should describe these procedures in detail, or get the information from someone who knows the equipment.
- When purchasing equipment, look for models that have been tested and endorsed for food safety by recognized agencies that certify products and write standards for food, water, air, and consumer goods.
- Three prominent agencies are
- These agencies are accredited by the American National Standards Institute (ANSI), as Standards Developing Organizations (SDOs).

- They are also accredited by ANSI to certify equipment (such as baking and other commercial food equipment) against the American National Standards that have been developed by each of these SDOs.
- The standards, and the certifications of these three agencies, are recognized internationally.
- Products meeting the testing requirements of these agencies are labelled or marked accordingly.
- Criteria govern such factors as design and construction (for example, sealed joints and seams, as well as accessible and component parts), materials used (for example, nontoxic materials, smooth and easily cleanable surfaces), and performance testing.

NSF International certification mark.	SANITATION NSF/ANSI STD. NO.	The Underwriters Laboratory logo.
www.nsf.org; formerly the National Sanitation Foundation	www.csa international.org	www.ul.com

Large Equipment: Mixers

Mixers, ovens, and dough-handling equipment take up most of this category

- Mixers of various types are essential tools in the bakeshop.
- Two main types of mixer are used in small and medium-size bakeshops: vertical and spiral.
- Other types of specialized equipment are used in large industrial bakeries.

Vertical Mixer

- Also called a planetary mixer, most common type used in baking.
- Planetary term is descriptive of the motion of the beater attachment. Just as a planet spins on its axis while revolving around the sun, so too does the beater attachment to reach all parts of the stationary bowl.
- Table top mixers range in capacity from 5 to 20 quarts (4.75 to 19 L)
- Floor models are available as large as 140 quarts (132 L)
- Be sure to use the right-size attachment for the bowl.
- Using a 40-quart paddle with a 30-quart bowl could cause serious damage to the equipment.
- The bowl and the mixing attachment should be firmly in place before turning on the machine.
- Always turn off the machine before scraping down the bowl or inserting a scraper, spoon, or hand into the bowl.

Mixer attachments

WHIP: Beating egg foams and creams

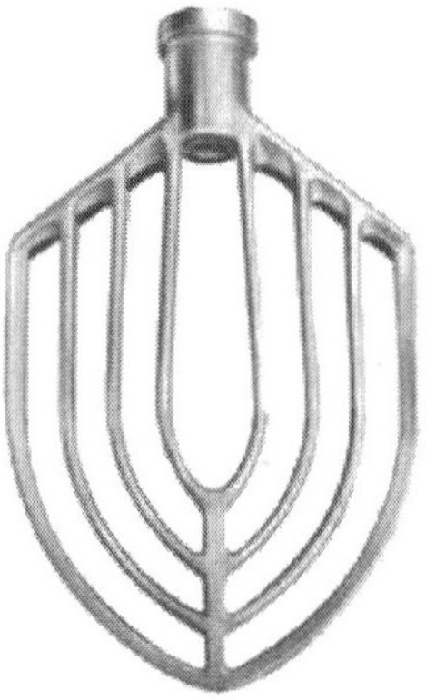

PADDLE: Flat blade for general mixing

DOUGH ARM/HOOK: Mixing and kneading yeast doughs, can be spiral or J-hook

Vertical Mixer

Additional special attachments are also available:

- The sweet dough arm combines the actions of the dough arm and the flat paddle and is used for mixing sweet doughs.
- The wing whip is used for mixing materials too heavy for the standard wire whip.
- The pastry blender is used to blend fat and flour, as in making pie doughs.
- The availability of such a variety of attachments points up one of the main advantages of the planetary mixer: its versatility.
- With a single machine, the baker can produce a great variety of doughs, batters, creams, meringues, and other goods.
- Vertical mixers have an attachment hub that can be used to power many other tools, such as grinders and slicers, also useful in kitchen

Small Table-Model Mixer.

Large Floor-Model Mixer.

Spiral Mixer

- Spiral mixers are designed for doughs and heavy batters
- Used primarily for making large quantities of yeast doughs for breads and bagels.
- Spiral mixers do not have interchangeable bowls and agitator arms.
- The agitator arm is in the shape of a spiral, and both the bowl and the spiral arm rotate to develop the dough quickly and efficiently.
- In a typical model, the bowl may be set to rotate in either direction.
- Dough capacities range from 5 to 30 pounds (2.3 to 14 kg) for small machines to more than 500 pounds (230 kg) in large machines.

Important Features and Advantages

- Spiral mixers blend and develop dough more efficiently than planetary mixers.
- They develop dough more intensively and in less time, resulting in lower machine friction and dough warming
- The design of the bowl and beater allows for a wide range of dough capacity for each machine.
- For example, a medium-size mixer may handle as little as 10 pounds (4.5 kg) of dough or as much as 200 pounds (90 kg).
- By contrast, a vertical mixer bowl handles only a narrow range of dough weights; too large or too small a quantity will not mix properly.
- They are sturdier and more rugged.
- They handle more dough, last longer, and require less repair and maintenance.

Spiral Mixer

Spiral Mixer: Varieties

- **Fixed-bowl mixers** - have a no removable bowl. The dough must be lifted out by hand.
- **Removable-bowl mixers** - have a bowl that may be removed from the machine, usually on a wheeled trolley. They are useful for high-volume operations, because a new bowl of ingredients may be wheeled into place as soon as the earlier batch is removed.
- **Tilt mixers** - the entire machine tilts up to deposit the finished dough onto a tray or another container.
- Most spiral mixers have two operating speeds, although some specialized machines for stiff doughs have only one speed.
- In a typical mixing procedure, the baker uses the first speed for the first phase of dough mixing, when the ingredients are blended, and switches to the second speed for the later phase of dough development.
- Machines with automated controls have timers to regulate each phase of mixing.
- Similar to spiral mixers are fork mixers. Instead of a single spiral agitator or beater, these machines have a two-pronged fork shaped beater that enters the bowl at about a 45-degree angle. Fork mixers are used specifically for bread doughs.

Fixed-Bowl Spiral Mixer

Removable-Bowl Spiral Mixer

Tilt Spiral Mixer

Continuous Mixer

- Small amounts of scaled ingredients enter the machine continuously at one end.
- The ingredients are blended and developed into a dough as they move through the machine.
- The finished dough emerges at the other end of the machine.
- These mixers are efficient because instead of having to blend, say, 200 pounds of flour into a dough all at once, they can take in the flour in 5-pound increments, making the blending much easier.

Continuous Mixer

Horizontal Mixer

- Large, industrial-size machines capable of handling as much as several thousand pounds of dough at a time.
- Each model is designed to work best with a specific range of products, such as bread doughs, pastry doughs, or soft doughs and batters.
- Beater or agitator designs differ for each of these specialized models.
- Many horizontal mixers are equipped with water jackets that surround the mixing container.
- Water of the desired temperature is circulated through the jacket, enabling the operator to control the dough temperature with great precision.

Horizontal Mixer

Large Equipment: Dough Handling

Dough Fermentation Trough

- This item is used to hold mixed yeast doughs during fermentation.
- Small operations might simply use large mixing bowls on stands instead.

Divider

Divider

- Dividers cut scaled pieces of dough into equal portions by means of a die or cutter attached to a hydraulic or mechanical lever assembly.
- For example, a divider may cut a 3-pound piece of dough (called a press) into 36 pieces, 1 1/3 ounces each, for making dinner rolls.
- After they are divided, the individual pieces must be rounded by hand

Divider-Rounder

- This machine divides the dough, as does a simple divider and it then automatically rounds the individual portions, greatly speeding makeup of the dough products.

Dough Sheeter

- A sheeter rolls out portions of dough into sheets of uniform thickness.
- It consists of a canvas conveyor belt that feeds the dough through a pair of rollers.
- To make thin sheets, the dough usually must be passed through the rollers several times.
- The operator decreases the space between the rollers after each pass.

Dough Sheeter

Divider - Rounder

Moulder

- A moulder rolls and forms pieces of bread dough for standard loaves, baguettes, and rolls, eliminating the need to perform these tasks by hand.

Moulder

Proofer

- A proofer is a special box in which the ideal conditions for fermenting yeast doughs can be created.
- The box maintains a pre-set warm temperature and humidity level appropriate to the specific dough.

Retarder-Proofer

- A combination of retarder and proofer
- A dough can be retarded for a pre-set time, after which the machine switches to proofing mode and warms up to a second pre-set temperature and humidity level. For example, breakfast breads can be made up the previous day, held, and be fully proofed and ready to bake when the shop opens the next morning.

Retarder-Proofer

Proofer

Large Equipment: Ovens

Oven

- Ovens are, the workhorses of the bakery and pastry shop.

- They are essential for producing breads, cakes, cookies, pastries, and other baked items.
- Ovens are enclosed spaces in which food is heated, usually by hot air (except in the case of microwave ovens, which are not especially useful in a bakeshop). Several kinds of oven are used in baking.
- Steam is important in baking many kinds of breads
- Ovens used in bakeshops, including deck ovens, rack ovens, and mechanical ovens, may have steam injected into them during part of the baking cycle.

Deck Oven

- Deck ovens are so called because the items to be baked—either on sheet pans or, in the case of some breads, freestanding—are placed directly on the bottom, or deck, of the oven.
- There are no racks for holding pans in deck ovens.
- Deck ovens are also called stack ovens because several may be stacked on top of one another.
- Breads baked directly on the floor of the oven rather than in pans are often called hearth breads, so another name for these ovens is hearth ovens.

Deck oven

- Deck ovens for baking bread are equipped with steam injectors.

Wood-Fired Oven

- Wood-fired brick ovens are similar in function to deck ovens in that items are baked directly on the oven floor.
- These ovens are used in some operations that produce artisan breads, as well as in some restaurants that serve pizzas and similar items.
- The heat is generated by a wood fire built inside the oven.
- The fire heats the thick brick floor and walls, which retain enough heat to bake foods.
- Gas-fired brick ovens are similar, but the heat in them is easier to control.

Rack Oven

- A large oven into which entire racks full of sheet pans can be wheeled for baking.
- Normal baker's racks hold 8 to 24 full-size sheet pans, but racks made specifically to go into rack ovens usually hold 15 to 20 pans.
- Rack ovens hold 1 to 4 of these racks at once.

- The ovens are also equipped with steam injectors.
- Although this usage is not strictly correct, you may hear the term rack oven used for conventional ovens, such as those found in restaurant ranges, because the pans are placed on racks rather than directly on the bottom, as in deck ovens.

Revolving Oven

Mechanical Oven

- In a mechanical oven, the food is in motion while it bakes.
- The most common type is a revolving oven, in which the mechanism is like that of a Ferris wheel.
- This mechanical action eliminates the problem of hot spots, or uneven baking, because the mechanism rotates the foods throughout the oven.
- Because of their size, mechanical ovens are especially useful in high-volume operations.
- Revolving ovens can be equipped with steam injectors.
- Each of the multiple trays in such an oven holds one or more sheet pans.
- The operator loads one tray at a time through the narrow door in the front.

Revolving Oven

Convection Oven

- Convection ovens contain fans that circulate the air and distribute the heat rapidly throughout the interior.
- The forced air makes foods cook more quickly at lower temperatures.
- However, the strong forced air can distort the shape of items made with batters and soft doughs, and the airflow may be strong enough to blow baking parchment off sheet pans.
- Therefore, convection ovens are not as versatile for the baker as are the other kinds of ovens discussed here.

Convection Oven

Steam-Jacketed Kettle

- Steam-jacketed kettles, or steam kettles, have double walls between which steam circulates.
- Liquids in the kettle itself are heated quickly and efficiently.
- Although restaurants may use large floor-mounted kettles for making stocks, smaller table models are more useful in the bakeshop for making custards, creams, and fillings.
- Tilting kettles with a pouring lip are called trunnion kettles.
- Table models range in capacity from a few quarts or litres to 40 quarts (38 L).

Steam-Jacketed Kettle

Fryer

- Fryers are needed in the bakeshop for doughnuts and other fried items.
- Small operations often use standard deep fryers (or even stovetop kettles), but larger doughnut fryers are best if you make doughnuts in quantity.
- They should be used in conjunction with screens, for lowering the doughnuts into the fat and removing them when fried.

Steam-Jacketed Kettle

- In the fryer in the illustration, the proofed doughnuts are arranged on the screen at the right side of the fry kettle.
- The operator then manually lowers the screen into the hot fat by means of the two raised handles.
- The illustration also shows, on the left side, a batter depositor for cake doughnuts.

Pans, Containers, Moulds

Many of the pots and pans found in the hot kitchen are also used in the bakeshop.

- Saucepans are used to boil syrups and to cook creams and fillings.
- Mixing bowls are used for mixing and various other activities
- Specialty containers and moulds for the bakery as follows:
- Moulds are of two types:
    - for baking dough or batter items,
    - for giving shape to refrigerated items such as Mousses and bombes

Cake Pans

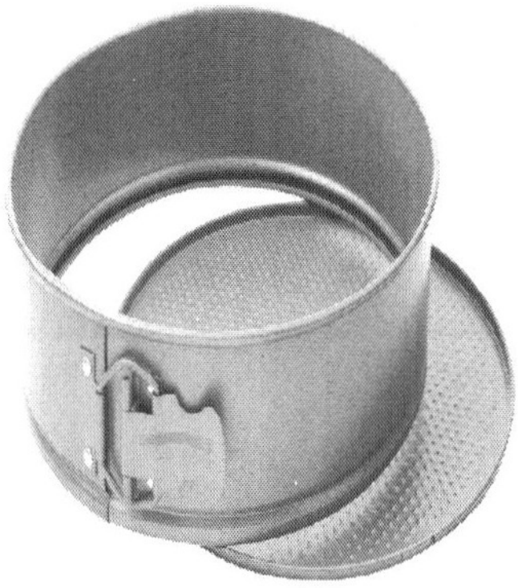

- Most cake pans are round, but other shapes, such as hearts, are available for specialty cakes.
- Cake pans come in many sizes.

Springform Pan

- A cake pan with a removable bottom.
- Used primarily for baking cheesecakes and other items too delicate to be easily and cleanly removed from standard cake pans.

Tube Pan

- A deep cake pan with a tube in the center.
- The tube promotes even baking of angel food cakes and similar items.

Muffin Pan

- Metal baking pan with cup-shaped indentations for baking muffins
- Pans are available for making muffins in several sizes.

Flexipan

- This is the brand name for a line of non-stick baking pans made of a flexible silicone material.
- Flexi pans are available in dozens of shapes and sizes to make a wide range of products, from muffins and quick bread loaves to petits fours.

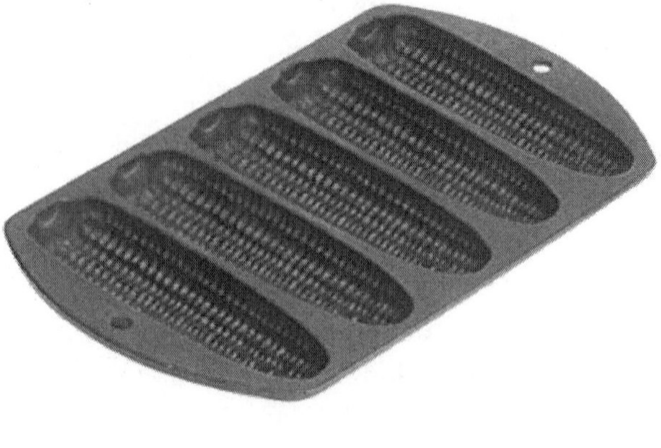

Cornstick Pan

- Special baking pan with indentations shaped like small ears of corn.
- Used for baking cornbread items.

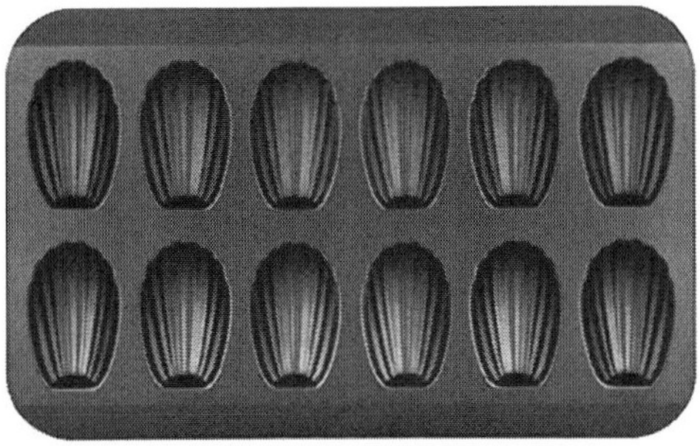

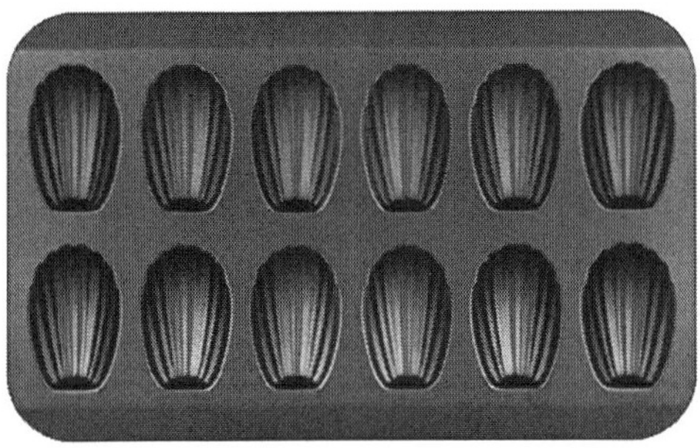

Madeleine Pan

- A special baking pan with shell-shaped indentations, used for baking madeleines

Hotel Pan

- A rectangular pan, usually made of stainless steel.
- Designed to hold foods in service counters.
- Also used for baking and steaming, and often for baked items such as bread pudding.

- The standard size is 10 × 20 inches (325 × 530 mm).
- Fractions of this size (1/2, 1/3, and so on) are also available.
- Standard depth is 2 1/2 inches (65 mm), but deeper pans are also available.

Loaf Pan

- A rectangular pan, usually with slightly flared sides, used for baking loaf breads.
- Loaf pans can also be used for moulding refrigerated and frozen desserts.
- A special type of loaf pan is the Pullman pan, which has straight, not flared sides, and a removable lid, for baking Pullman loaves of bread

Sheet Pan

- A shallow, rectangular pan (1 inch/25 mm deep) for baking sheet cakes, cookies, rolls, and other baked goods.
- A full sheet pan measures 18 × 26 inches (46 × 66 cm).

- Half-sheet pans are 13 × 18 inches (33 × 46 cm).
- Perforated sheet pans are the same size, but the bottom is full of tiny holes.
- These allow even baking and browning of breads and rolls because the holes let the oven's hot air circulate freely around the items as they bake.

- Pan extenders are metal or fiberglass frames that fit inside sheet pans.
- They give straight sides to sheet cakes and make the pan deeper.
- Extenders are usually 2 inches (5 cm) high.

Pie Pan

- Shallow pan with sloping sides, used for baking pies.
- Disposable aluminium pie pans are usually used in retail bakeshops.

Tart Pan

- A shallow (1 inch/2.5 cm deep) metal pan, usually with fluted sides, used for baking tarts.
- Standard pans are round, but square and rectangular pans are also available.
- They may be made in 1 piece or with removable bottom to make removal of baked tart from pan easier.
- Tart pans make multi serving pastries, but smaller tartlet pans make single-portion tartlets.
- Like tart pans, these come in a variety of sizes.
- The smallest usually are in one piece and lack the removable bottom.

Mixing Bowls

- The most useful mixing bowls are made of stainless steel and have round bottoms.
- They are used for general mixing and whipping.
- The round construction enables the whip to reach all areas, for thorough mixing or whipping.

Baba Mould

A small thimble-shaped mould for making babas

Banneton

- A bentwood basket, available in various shapes, for holding and giving shape to certain hearth bread doughs as they proof.
- Similar canvas-lined baskets are also available.

Barquette

- A small boat-shaped mould for petits fours and small tartlets.

Bombe Mould

- A dome-shaped mould for frozen desserts

Brioche Mould

- A flared pan with fluted sides for making brioche

Cake Ring/ Charlotte Ring

- Also called cake rings, these are stainless-steel rings in various diameters and heights, most often used for making moulded desserts and for shaping and holding desserts made of layers of cake, pastry, and fillings.
- The rings are removed after the fillings have set and before serving or display.

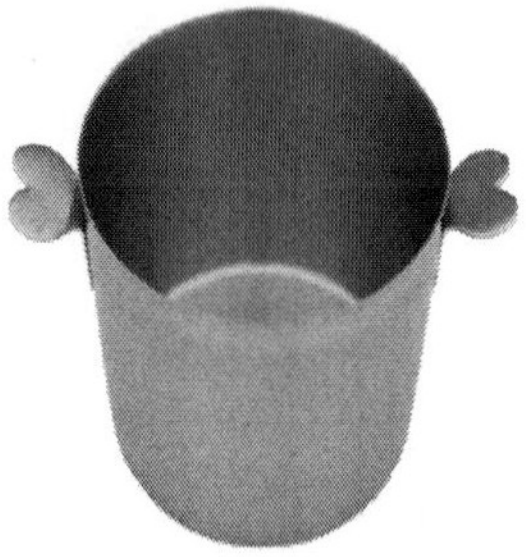

Charlotte mould

- The classic charlotte mould is round, tapered, and flat-bottomed, with two handles near the top rim.
- Except for the Apple Charlotte, which is baked in this mould, classic charlottes are made with a Bavarian cream filling and refrigerated until set, not baked.

Chocolate Moulds

- Used for all sorts of chocolate work, from large display pieces to bite-size truffles.

Petit Four Moulds

Tiny metal moulds in a variety of shapes, used for baking an assortment of little tartlets, financiers and other petits fours.

Savarin Mould

Small ring-shaped or doughnut-shaped metal mould for baking savarins

Hand Tools

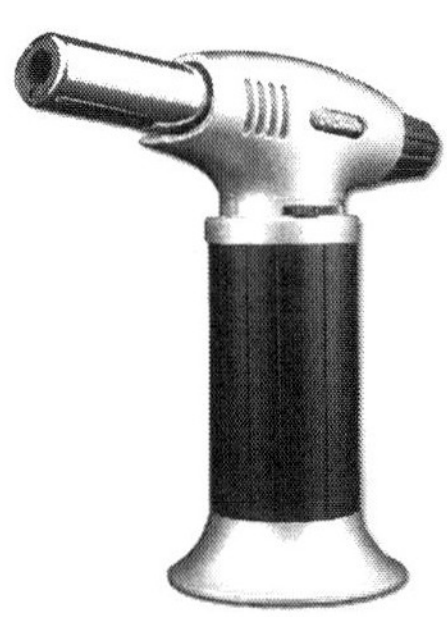

Blowtorch

- A tool used for caramelizing and controlled browning of various pastry items, and for caramelizing the sugar topping of creme brulee.
- Butane or propane is used as fuel, depending on the model.

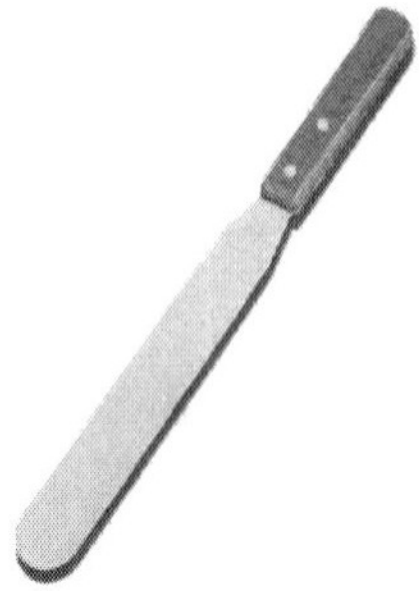

Bowl Knife

- Also called a straight spatula or palette knife, this tool has a long, flexible blade with a rounded end.
- Used mostly for spreading icing on cakes and for mixing and bowl scraping.
- A variant with an angled blade is called an offset spatula.
- The bent blade allows spreading and smoothing batters and fillings inside pans.

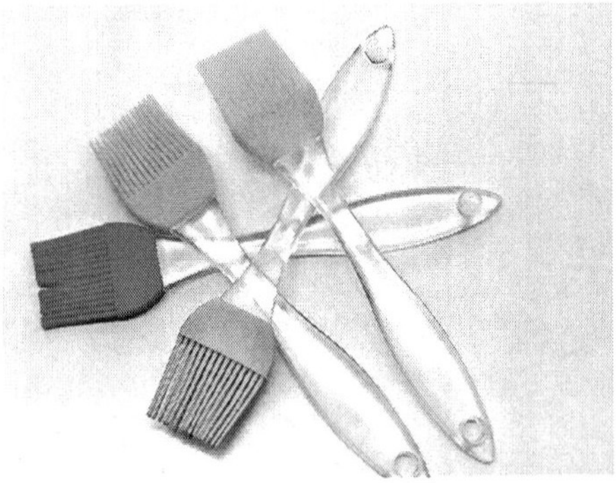

Brushes

- Pastry brushes are used to brush items with egg wash, glaze, and so on. Larger bench brushes are used to brush flour from table tops and from the surface of dough.
- Oven brushes are used to clean excess flour from deck ovens

Chinois and China Cap

- A chinois is a conical strainer with a fine mesh, used mostly for straining sauces.
- A china cap is also a conical strainer, but it is made of perforated steel, so it doesn't strain as finely.
- A china cap is usually lined with several layers of cheesecloth if the liquid must be well strained.

Icing Comb

- A small plastic tool, usually triangular, with serrated edges in various patterns, for decorating icings and other pastry and decorative items.

Cutters

- Many types of cutters are used in the pastry department.
- Cookie cutters and pastry cutters, available in many shapes, cut decorative shapes by stamping them from rolled-out dough.
- Roller cutters have a handle on each end, like a rolling pin, and are rolled over rolled-out dough to cut repetitive shapes quickly and efficiently, with minimal loss of dough to trimmings and scraps.
- Roller cutters are often used for croissants

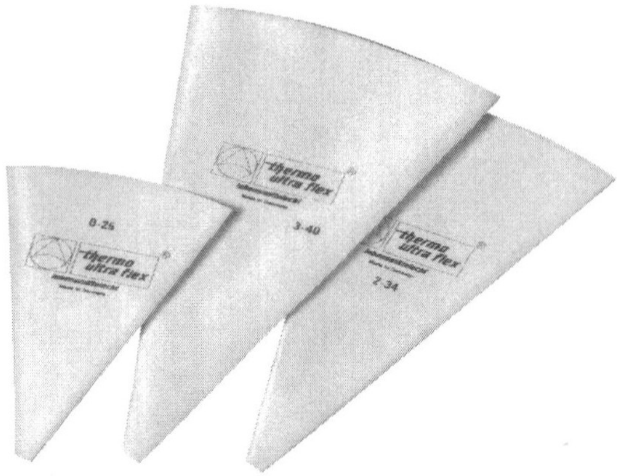

Pastry Bag

- A cone-shaped cloth or plastic bag with an open end that can be fitted with metal or plastic tubes or tips of various shapes and sizes.
- Used for shaping and decorating with items such as icing; for filling certain kinds of pastries and other items, such as eclairs; and for portioning creams, fillings, and doughs.

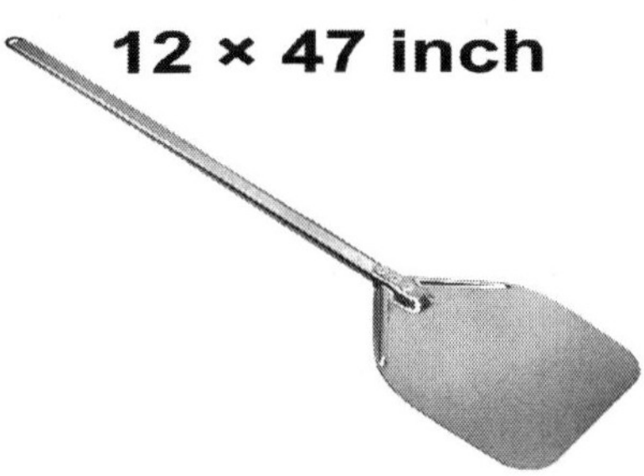

Peel

- A thin, flat wooden board or steel sheet with a long handle, used for inserting and removing hearth breads from deck ovens.
- As they are thinner than traditional wooden peels, steel peels are easier to slide under baked loaves.

Roller Docker

- A tool that pierces holes in rolled-out dough to prevent bubbling during baking.
- It consists of a handle attached to a rotating tube fitted with rows of spikes.

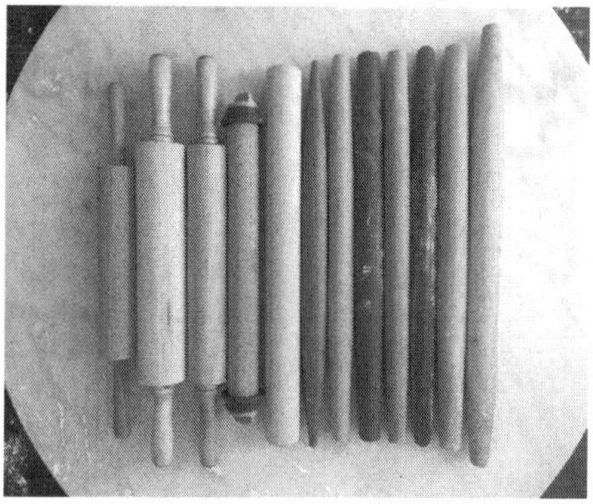

Rolling Pins

- Many types of rolling pin are used in the bakeshop for rolling out doughs.
- Perhaps the most versatile pin, used for most general rolling tasks, is simply a solid hardwood rod, about 2 inches (5 cm) thick and 20 inches (50 cm) long.
- A French rolling pin is about 2 inches (5 cm) thick at the center and tapered toward the ends.
- It is useful for rolling pie doughs and other doughs that must be rolled to a circular shape.
- For large quantities or heavy work, a heavy ball-bearing pin may be used.
- This pin is 3 to 4 inches (8 to 10 cm) thick and has a swivelling rod inserted through the center, with a handle at each end.
- Textured rolling pins are used to emboss designs, such as a basket weave pattern, in sheets of marzipan and pastillage.

Scrapers

- A bench scraper, also called a dough scraper, is a small rectangle of stainless steel with a handle along one of the long edges.

- It is used for cutting and portioning dough and for scraping table tops.
- A bowl scraper is a piece of plastic about the same size, but with one curved edge and no handle.
- It is used for scraping out the contents of mixing bowls.

Sieve

- A round metal screen supported in a stainless-steel hoop frame. It is used for sifting flour and other dry ingredients. Also called a drum sieve or tamis (pronounced tah-mee).

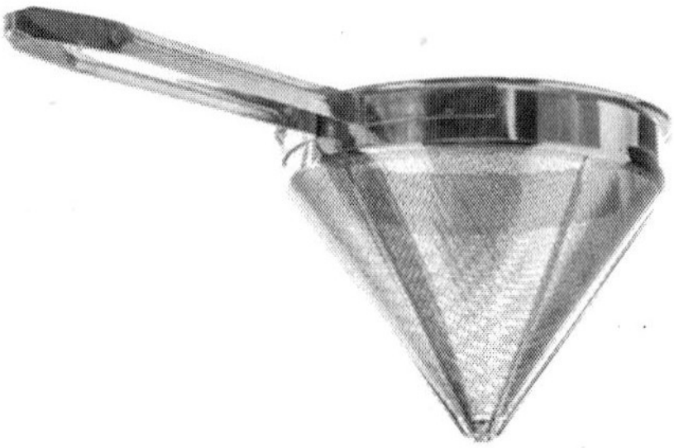

Strainer

- A round-bottomed, cup-shaped tool made of screen mesh or perforated metal, with a handle on one side.
- Used for separating solids from liquids, such as draining the juice from fruit.
- Screen-mesh strainers can also be used for sifting dry ingredients, like a sieve.

Turntable

- A round, flat disk that swivels freely on a pedestal base.
- Used for holding cakes for decorating.

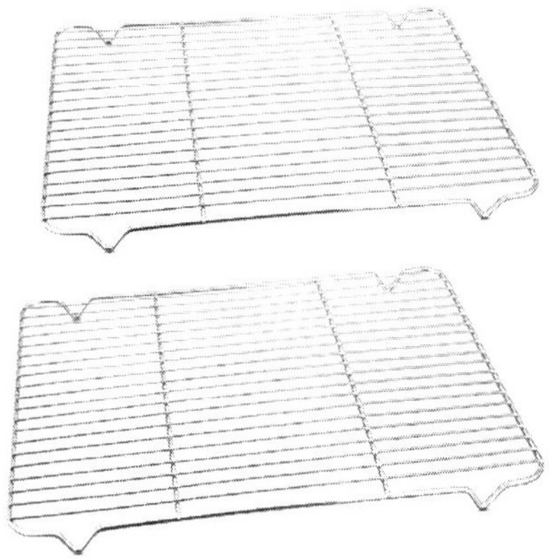

Wire Rack

- A wire grate used to hold baked goods as they are cooling, or to hold items such as cakes while liquid icings, such as fondant, are applied.

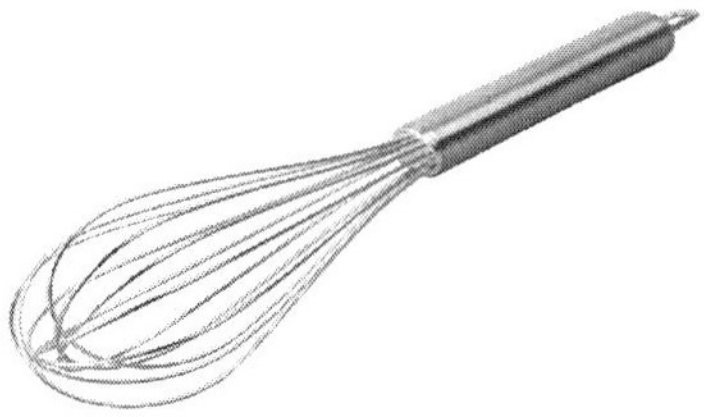

Whisk

- Loops of stainless-steel wire fastened to a handle.
- Whips with a few stiff wires are used for mixing and blending, and whips with many flexible wires are used for whipping foams, such as whipped cream and egg foams. Also called whisk.

Miscellaneous Tools and Equipment

Acetate

- A type of clear plastic.
- Acetate strips are used for lining charlotte moulds in the production of certain cakes, pastries, and refrigerated desserts.
- For retail display, the strips can be left on after the charlotte rings are removed to support the dessert while displaying the layers.
- Acetate sheets are most often used in decorative chocolate work

Couche

- A sheet of heavy linen or canvas, used for supporting certain breads, such as baguettes, as they are proofed.
- The cloth is placed on a sheet pan and pleated to form troughs to hold the loaves so they can proof without spreading.

Silicone Mat

- Flexible fiberglass mat coated with non-stick silicone, used to line baking sheets.
- Available to fit full and half-size sheet pans.
- Also used in sugar work
- The mats withstand temperatures up to about 480°F (250°C) and can be reused indefinitely if well cared for and not folded or creased.
- There are several manufacturers of silicone mats, but they are often known by one brand name, Silpat.

Ice Cream Freezer

- Machine for churning and freezing ice creams and sorbets.
- It consists of a large refrigerated canister or container with a paddle, called a dasher, that rotates inside.
- The ice cream or sorbet mix freezes against the walls of the canister and is continually scraped off and remixed to prevent the formation of ice crystals.
- Unlike home models, which depend on a salted ice water mixture to create freezing temperatures, commercial ice cream freezers contain a built-in electrically operated freezing unit.

Marble Top Table

- A stone material used for table top or work surfaces in pastry shops.
- The hard, cool surface of marble is ideal for working with pastry doughs, as well as for tempering chocolate and for some decorative work, such as pastillage.
- Marble slabs may be installed on top of under-the-counter refrigerated storage boxes.
- This keeps the marble cool even in warm weather.

Parchament Paper

- Also called baking paper or silicone paper, this is a sheet of treated non-stick paper sized to fit standard sheet pans.
- When used to line pans, parchment eliminates the need for greasing them.
- Also used to make piping cones for decorative work

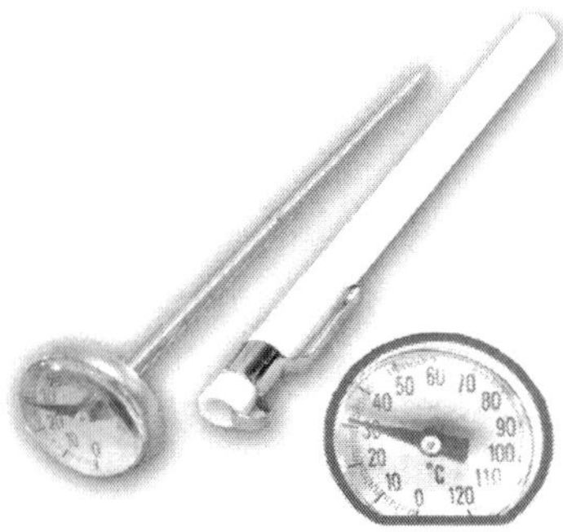

Thermometers

- Thermometers have many uses in the bakery, and there are many types of specialized thermometers.
- The sugar thermometer, also called a candy thermometer, is one of the most important.
- It is used for measuring the temperature, and hence the concentration, of boiling sugar syrups
- The chocolate thermometer is used for tempering chocolate.
- Other thermometers measure the temperature of bread doughs, frying fat, and the interiors of ovens, refrigerators, and freezers (to check the accuracy of the equipment's thermostat).

Hydrometer

- Also called a sugar densimeter, saccharometer, and Baumé hydrometer.
- Used to test the density of sugar syrups. (Sometimes called a thermometer, but this is inaccurate because it doesn't measure temperature.)
- It is a glass tube, weighted at one end, that is floated in the solution to be tested.
- Because it floats higher in denser solutions, the density can be read off the scale marked along the length of the tube at the point where the surface of the liquid meets the tube.

2. COMMODITIES (INGREDIENTS USED IN BAKERY AND CONFECTIONERY)

WHEAT

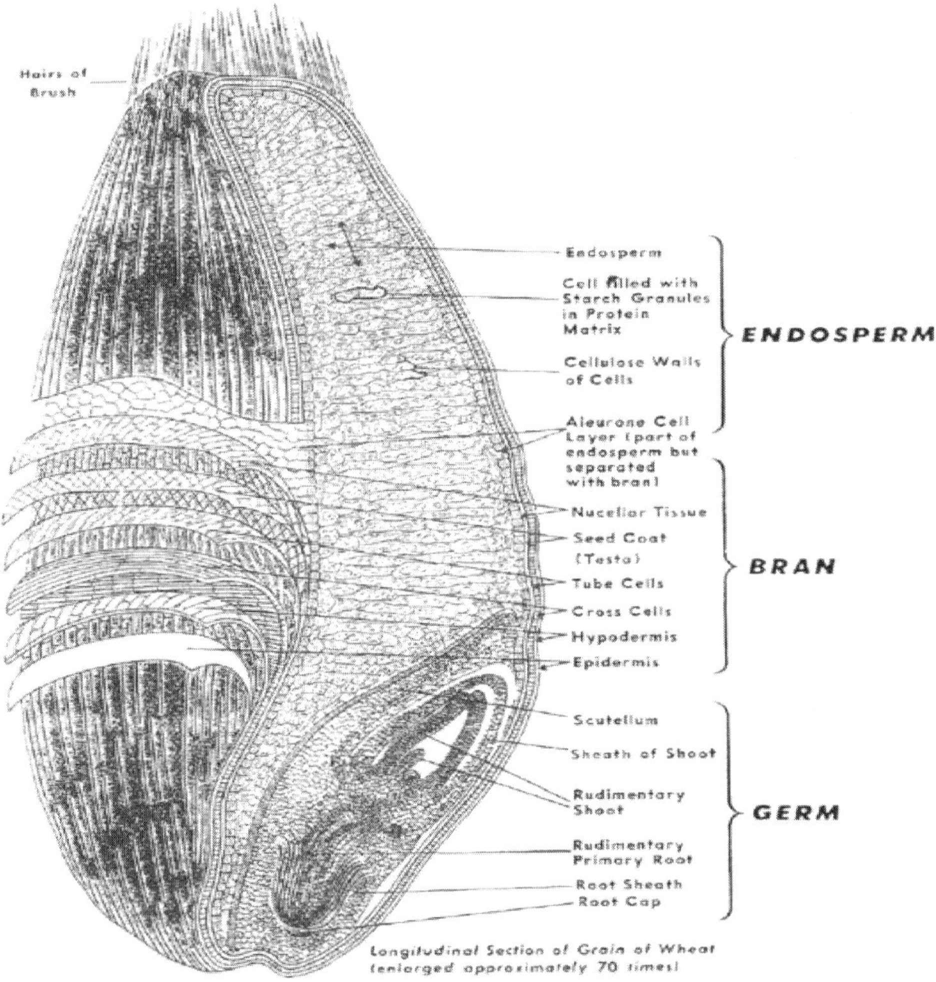

Longitudinal Section of Grain of Wheat (enlarged approximately 70 times)

Wheat contains huge measure of gluten, Gluten is the versatile clingy, extreme substance framed from the insoluble proteins of wheat flour during mixture advancement. It is the gluten which permits development of air cells inside the mixture as it warms. This is fundamental to a raised item. Subsequently, wheat improves bread then different grains. The main other grain which contains obvious measure of gluten is rye.

Wheat flour is the most important ingredient in the bakeshop. It gives mass and structure to the vast majority of the pastry specialist's items, including breads, cakes, treats, and baked goods. Not at all like the home cook who depends on the whole on an item called generally useful flour, the expert bread cook approaches a wide assortment of flours with various characteristics and attributes. To choose the best possible flour for every item, and to deal with each accurately, you have to comprehend the attributes of each sort of flour and how it is processed.

Composition of Wheat

The wheat kernel consists of three main parts as below:

Bran

Germ

The germ is the part of the kernel that becomes the new wheat plant if the kernel is sprouted. It has a high fat content that can quickly become rancid. Therefore, whole wheat flour containing the germ has poor keeping qualities. Wheat germ is high in nutrients, containing protein, vitamins, and minerals, as well as fat.

Endosperm

The endosperm is the white, starchy part of the kernel that remains when the bran and germ are removed. This is the portion of the wheat kernel that is milled into white flour. Depending on its source, the wheat endosperm contains about 68 to 76% starch and 6 to 18% protein. The endosperm also contains small amounts of moisture, fat, sugar, minerals, and other components. These are discussed in more detail when we consider the composition of flour in a later section. are a much lighter, creamy white color.) Bran is high in dietary fiber and contains B vitamins, fat, protein, and minerals

Flour Grades

Different grades of flour come from different portions of the endosperm. Modern milling processes were developed to separate these portions.

Patent Flour

Flour from the interior of the endosperm, extracted during the first streams of milling, is considered the highest grade of flour and is called patent flour. It is fine in texture because the granules from the interior of the grain are the smallest. It is also whiter in color than other grades and has high-quality protein. It is nearly completely free of any trace of bran or germ.

Different grades of patent flour are available, depending on the amount of endosperm extracted Fancy patent, also called extra short, is made from only the inner 40 to 60% of the endosperm.

Short patent may contain up to 80% of the endosperm, while long patent consists of up to 95% of the endosperm. Most bakers use the term patent flour to mean strong patent flour used for breads.

However, any flour made from the interior of the endosperm is patent flour, even if made from soft wheat. Cake flour and pastry flour are also patent flours. You should be aware of these two uses of the term to avoid confusion.

Clear Flour

The portion of the endosperm left after the patent flour has been removed is called clear flour.This flour comes from the outer parts of the endosperm and thus is darker in color and higher in protein. Clear flour is usually separated into more than one grade. First clear is dark flour, tan in color that is often used in rye breads, where its dark color is not noticed, and its high protein content contributes much-needed gluten. Even though it is dark, first clear is lighter in color than second clear, which is low-grade flour not usually used in food production. The outer parts of the endosperm are higher in protein than the interior. Thus, for each type of wheat, clear flour is higher in protein than patent flour. However, the quality of the protein in patent flour is better. This means that gluten that is formed from these proteins stretches well and makes a strong, elastic film.

STRAIGHT FLOUR

Straight flour is made by combining all the streams of the milling process. In other words, it is made from the entire endosperm. Because it contains the darker parts of the grain as well as the whiter interior, straight flour is darker in color than patent flour. In addition, it contains small amounts of bran and germ that weren't separated during milling. Straight flour is not often used in North American baking. Some European flours are straight flours.

Extraction

Extraction refers to the amount of flour milled from a given amount of grain. It is expressed as a percentage of the total amount of grain. For example, whole wheat flour is said to be 100% extraction because if you start with 100 pounds of grain, you end up with 100 pounds of whole wheat flour. As a second example, if a grade of flour is described as 60% extraction, this means It would take 100 kilograms whole grain to produce 60 kilograms of this grade of flour. The remaining 40% is bran, germ, shorts, and darker, lower grades of flour. Patent flour is low extraction flour, while straight flour is high-extraction flour.

Atta

Atta flour is probably the most commonly used wheat flour in the Indian subcontinent. It is a whole- wheat flour made from a durum wheat, which is a very hard wheat (similar to the flour used to make pasta). Durum wheat has a high protein content and a high gluten content; this makes it ideal for making breads. Atta also has a high bran content, which is very healthy.

Traditionally Atta is stoneground and comes out a sort of creamy brown color. It is quite a coarse flour but very tough so you can roll it out very thin. It is used to make roti, chapatti, parathas, and quite a few other Indian breads. It is sometimes marketed in the west (and referred to in recipes) as chapatti flour.

Maida

Maida is softer general-purpose flour (sometimes called 'all-purpose flour') and is a highly refined flour. In the west, you could substitute plain white flour. It can be used for making roti and parathas and this gives a much softer result than the traditional Atta flour, but it is necessary for making naan bread and samosas. I usually use this for pooris as well. In India, Maida is always used to make pastries and sweeter breads such as sheermal, a sweet milk bread, or baqerkhani, a layered bread like rough puff pastry - both from Kashmir.

Semolina

Semolina (sooji) is made from hard durum wheat. It is the starchy endosperm part only which is separated from the bran and the wheat germ and then milled into flour. This is not just an Indian flour but has many uses all over the world; it is used to make pasta in Italy and semolina pudding in English schools.

In India, it is used for making rava dosa (a quick version of a dosa that does not need fermentation). I have used it mixed in with Atta flour to make pooris. It is also used for making sooji halva and Indian sweet made from semolina and ghee, and then flavored with nuts or fruit or coconut etc.

Lapsi Fada

This is more like cracked wheat - it is a whole-wheat flour that is only partly ground. It is used to make sweet dishes for special occasions like marriages.

Rice Flour

Rice flour is made from ground rice kernels and is sometimes called ground rice, it can be made from white or brown rice with brown rice flour having a greater nutritional value. It is easier to digest than wheat flour and so often finds its way into baby food, it is naturally gluten free it is a good substitute for those who are gluten intolerant.

In the south of India, it is used to make various pancakes and breads; idlis and dosas are two such examples - made from a mixture of rice and lentil flours. Karnataka a state on the southwestern coast makes a roti from rice flour called akki roti. Sweets similar to halva called shondesh or malpua are made from a mixture of wheat and rice flour with sugar and coconut.

It is used as a thickening agent and is particularly good for thickening products which are to be frozen since it stops the liquid content from separating out. In addition, rice flour is used to bulk out other products such as ground spices.

Outside India, it is used for making rice noodles, crackers, batters, cereals and cakes amongst many other things.

Dhal Flours

There are as many types of dhal flour as there are type of dhal. They are generally gluten free, also have low glycemic indices, and so are very good for people with certain dietary requirements. The main two are gram flour and urad dhal flour, but there are also flours made from moong beans and toor dhal and masoor dhal.

Gram Flour

A very common Indian flour, also called besan and chickpea flour this is made from dried chickpeas. It has a very earthy smell, a slightly nutty taste and is very high in protein.

Probably it is most well-known use is to make the batter for onion bhajis and other pakora but it is also used as a thickener in curries. It is used to make fried batter balls called boondi, these can then be dipped in syrup to make laddu. In northern India it is used to make a yoghurt-based soup called kadi - yoghurt, gram flour and spices.

Urad Dhal Flour

This is a real Indian flour, not used much outside Indian cuisine, also called black lentil flour and papad flour it is used to make poppadums' and of course to mix with rice flour to make idlis and dosas. This flour is also one of the most popular for making vadai - deep fried doughnuts make in the South of India.

Other Flours, Meals, And Starches

Rye Flour

Next to white and whole wheat, rye is the most popular flour for bread making. Although rye flour contains some proteins, these do not form gluten of good quality. This is because although it contains enough gliadin, rye flour does not contain enough glutenin. Therefore, breads made with 100% rye flour are heavy and dense. To make a lighter rye loaf, it is necessary to use a mixture of rye and hard wheat flours. Typical formulas call for 25 to 40% rye flour and 60 to 75% hard wheat flour. Rye flour is also high in pentosan gums—about four times as much as wheat flour. The gums give some structure to rye breads, but they also interfere with gluten development and make rye doughs stickier than wheat doughs.

The following **grades** and **type** are generally available:

- **Light rye.** The lightest is nearly white. It has a very fine texture and a high percentage of starch, with little protein.
- **Medium rye.** This is straight flour, milled from the whole rye grain after the bran is removed. Thus, it is darker than light rye and has higher protein content.
- **Dark rye.** Like clear flour milled from wheat, dark rye comes from the part of the rye grain closest to the bran. Thus, it is darker than other rye flours and has a lower percentage of fine starch particles.
- **Whole rye flour.** This product is made from the whole rye kernel, including the bran and germ.
- **Rye meal or pumpernickel flour.** Rye meal is a dark, coarse meal made from the entire rye grain, including the bran and germ. Products labeled pumpernickel are sometimes cut into flakes rather than ground into coarse meal. Rye meal is used for pumpernickel bread and similar specialty products.
- **Rye blend.** This is a mixture of rye flour (generally 25 to 40%) and strong wheat flour, such as clear flo

Corn

Wheat and rye account for the great majority of the grain flours and meals used in the bakeshop.Other grains are used mainly to add variety to baked goods.Of these other grains, corn is perhaps the most important.(Note: In Great Britain, corn is referred to as maize, while the word corn simply means "grain.") Corn contains no gluten-forming proteins, although it does contain significant quantities of other proteins, and is therefore important in vegetarian diets. Corn is most often used by the baker in the form of yellow cornmeal.

Cornmeal is available in grinds from fine to coarse. Coarse cornmeal produces a crumbly, somewhat gritty texture in corn breads, a quality that is desirable in some products.

Spelt

Spelt is considered an ancestor of modern wheat. Like wheat, it contains gluten proteins, but they form a rather weak gluten structure that can't withstand much mixing. Spelt has a lower absorption ratio than wheat. Spelt was unheard of by most bakers until not long ago. More recently, it has enjoyed increase popularity, partly because of increased interest in vegetarian diets and the desire for greater variety in dietary sources of protein. It is found increasingly as an ingredient in specialty breads.

Oats

Long familiar as breakfast porridge, oats in various forms also find uses in the bakeshop. Although rich in protein, including enough gluten proteins to make them off-limits for people allergic to gluten, oats do not form a gluten structure when mixed into dough. Oats are high in gums, which supply dietary fiber.

The gum content accounts for the gummy or gluey texture of oatmeal porridge.

- **Rolled oats**, commonly used for porridge, are made by steaming oat grains to soften them and then flattening them between rollers.
- **Steel-cut oats** are whole grains that have been cut into small pieces. They are occasionally used in small quantities in specialty breads.
- **Oat flour** is whole-grain oats ground into fine flour, which can be mixed with wheat flour in small quantities for specialty breads.
- **Oat bran** is a good source of dietary fiber that is often used as a muffin ingredient.

Buckwheat

Buckwheat is technically not a grain because it is the seed not of a grass but of a plant with branched stems and broad, arrow-shaped leaves. Whole buckwheat is often ground into dark, strong-tasting flour, while buckwheat endosperm alone is ground into lighter-colored flour with a somewhat milder taste.

When the grains are crushed into small pieces, they are called buckwheat groats and can be cooked like rice. Buckwheat flour is most commonly used for pancakes and crêpes, but it can also be used in small quantities in specialty breads and multigrain products.

Soy

Soy is not a grain; it is a bean, or legume. Nevertheless, it may be ground into a flour like a grain. Unlike regular grains, however, it is low in starch. It is also high in fat and protein, although it contains no gluten proteins. The rich protein content makes it valuable in vegetarian diets. Raw soy flour contains enzymes that make it useful in baking. These enzymes aid yeast action and bleach the pigments in wheat flour. Raw or untoasted soy flour should be used in small quantities in yeast breads, generally about 0.5%. Higher quantities give an unpleasant beany flavor to breads and produce poor texture.

When soy flour is toasted, the enzymes are destroyed, and the flour has a pleasanter flavor.

Rice

Rice flour is smooth white flour milled from white rice. It has a small amount of protein but no gluten, so it is often used in gluten-free baked products.

Other Grains and Flours

Many other grains, such as amaranth, millet, teff, and barley, have limited used in the bakeshop, either as flour or as whole grains. Other starchy nongrain foods, such as potatoes and chestnuts, can be dried and ground into flour for special products. Cooked potato starch is sometimes added to yeast breads, because the starch is easily broken down by diastase enzymes into forms of sugar that yeast can use.

Starches

In addition to flours, other starch products are used in the bakeshop. Unlike flour, they are used primarily to thicken puddings, pie fillings, and similar products.

The three most important starches in dessert production are as follows:

1. **Corn-starch** has a special property that makes it valuable for certain purposes. Products thickened with corn-starch set up almost like gelatine when cooled. For this reason, corn-starch is used to thicken cream pies and other products that must hold their shape.
2. **Waxy maize** is made from a different type of corn. It is almost always manufactured into a form called modified food starch. Waxy maize and other modified starches have valuable properties. Because they do not break down when frozen, they are used for products that are to be frozen. Also, they are clear when cooked and give a brilliant, clear appearance to fruit pie fillings. Waxy maize does not set up firm like corn-starch but rather makes a soft paste that has the same consistency hot and cold. Thus, it is not suitable for cream pie fillings.
3. **Instant starches** are precooked or pregelatinized, so they thicken cold liquids without further cooking. They are useful when heat will damage the flavour of the product, as in fresh fruit glazes such as strawberry.

Amylose and Amylopectin Starches

Starch molecules fall into two principal categories. Amylose molecules are long, straight chains, while amylopectin molecules have many branches. Most grain starches are high in amylose starches; starches from roots and tubers, such as potatoes and arrowroot, are high in amylopectin starches. The differences between the two types can be summarized as follows:

- Amylose starches, after cooking, get thicker and cloudier as they cool.
- They form a firm gel when cooled.

- They tend to break down and release liquid after long storage or after freezing.
- Amylopectin starches do not get thicker as they cool, and they remain fairly clear.

pH Value of Flour

pH

Also known as measure of Acidity and Alkalinity

pH is a key quality indicator in the food industry, influencing things such as shelf life and quality. It stands for "potential of hydrogen," and is the degree (intensity) of the acidity or alkalinity of a given product.1

In the baking industry, pH helps food scientists and technologists know how to better:

- mix different types of dry and wet ingredients
- balance flavours
- determine the exact final point of a given formulation and/or process

This concept should not be confused with the amount of acid (titratable acidity) present in the food

pH works based on the fact that molecules (acids and bases in aqueous solutions) disperse throughout the liquid, split up (dissociate), and expose their free hydronium [H_3O_+] and hydroxide ions [OH−].1

It is represented on a scale from 0 to 14. A pH value of 0 corresponds to an extremely acidic product, and a value of 14 corresponds to an extremely alkaline product. Most bakery products range from 3.5 to 8.0.

Application

The acidity or pH of foods is important for several reasons:

Certain pathogenic bacteria (those responsible for foodborne illnesses) cannot survive in acidic environments. In such cases, acidity plays a vital role as a preservation or inhibition mechanism, as in the case of sourdoughs.

Low pH values in products (4.0–5.5) inhibit the growth of bacteria and mold. For example, rope (Bacillus subtilis spores) in doughs or breads cannot grow with values between 5.4 and 5.5.

Such low pH can be obtained by using pH control agents (acidulants) such as propionic acid (commercially marketed as calcium propionate). A mold inhibitor, calcium propionate is most effective below pH 5.5. At pH lower than 5.5 (excess H+ concentration), the potentially active component, propionic acid, is undissociated, and becomes active in excess of H+ ions.1

pH can also be modified by using vinegar (acetic acid), which is another example of both a pH control agent and a preservative (also acting against rope contamination). It has little intrinsic antimicrobial activity, and so is added to increase the acidity (reduce the pH) and retard the initial growth of the bacteria.3

During fermentation (alcoholic and/or lactic), there is an increase in dough acidity (lower levels), which contributes to the shelf life, flavor, color, and rheological properties of the product.

Neutralization of baking powders (the amount in grams of sodium bicarbonate that are needed to neutralize 100 grams of an acid leavener to convert the bicarbonate to carbon dioxide). In this case, pH is an indicator of the extent of such a reaction, and results in a final neutral value (7.0).4

In cases where the acid leavener does not neutralize the sodium bicarbonate (baking soda), then the excess bicarbonate will increase the degree of browning of the sugars during baking. Maillard browning increases at higher pH values (>8.0). To prevent this, additional acid, e.g., cream of tartar (potassium bitartrate) is added to lower the pH and neutralize remaining baking soda.4

In gluten-free applications, where addition of hydrocolloids is a must, the pH of the product can be critical in terms of product stability (gelling). A hydrocolloid molecule held at its isoelectric point, i.e., the pH at which there is zero net charge, will likely come out of solution as it becomes insoluble in water.5

Measurement of pH

The pH value of any product can be measured by placing a pH meter device in direct contact with the product (previously prepared), or can be obtained quantitatively by following a math expression.

A pH meter determines electric potential between glass and reference electrodes, using a commercially calibrated apparatus standardized against known buffer solutions (a method known as potentiometry). The basic principle of potentiometry involves the use of an electrolytic cell composed of two electrodes immersed into a test solution. A voltage develops, which is related to the ionic concentration of the solution.

pH measurement of a product sample can be performed by following these steps:

1. Prepare 100 mL of deionized water.
2. Place 15 g of product sample into a blender with the water.
3. Thoroughly blend mixture for 10–30 seconds.
4. Pour it into a beaker to measure the pH with the potentiometer (pH meter).

Calculation of pH2

pH calculation can be done with the following equation:

$$pH = -\log10\,[H3O+]$$

[ ] means molar concentration of ions (in moles per liter or Molar).

The pH scale6

pH is represented on a scale from 0 to 14.

Acidic product	pH < 7.0	[H3O+] > 1.0 X 10-7 M
Neutral product	pH = 7.0	[H3O+] = 1.0 X 10-7 M
Basic, or alkaline, product	pH > 7.0	[H3O+] < 1.0 X 10-7 M

The neutral point, pH 7.0 (value of pure water), is the midpoint of the pH scale. A pH of 0 indicates extreme acidity, and a pH of 14 extreme alkalinity.

pH is a logarithmic value. H3O+ molar concentrations progress arithmetically, but each unit increment of pH indicates a tenfold increment in H3O+ concentration.

As a consequence, successive increments of pH do not indicate the same increments of H3O+ concentration. Therefore, from pH 7 to 6 there is a tenfold increase in acidity; from 7 to 5, the increase is 100-fold; and from 7 to 4, it is 1000-fold.6

Approved methods for pH measurement

- AOAC 981.12 (pH meter – matrices: vegetables/acidified foods)
- AACC 02-52.01 (Electrometric method – matrices: flour and some bread, crackers, cakes, and pastry products)

pH values of some articles

Food Product	pH
Baker's yeast bread	5.3–5.8
Sourdough bread	3.8–4.6
Wafer sheets	6.8–7.4
Snack crackers	5.5–5.6
Soda crackers	7.2-8.0
Most biscuits	7.0–7.2
Eggs	6.6
Honey	3.7–4.2
Milk (cow)	6.4–6.8
Miscellaneous	pH
1 M HCl solution	0.0
Vinegar	2.8
Apple juice	3.8
Water (pure)	7.0
Sea water	8.5
Milk of Magnesia	10.5
Ammonia	11.0
Bleach	12.0
1 M NaOH solution (lye)	14.0

References

Cauvain, S.P. "Testing Methods." Baking Problems Solved, 2nd ed., Woodhead Publishing Limited, 2017, pp. 469–470.

Nielsen, S.S. "Standard Solutions and Titratable Acidity." Food Analysis Laboratory Manual, 3rd ed., Springer International Publishing, 2017, pp. 179–184.

Cauvain, S.P. "Other Ingredients." Baking Problems Solved, 2nd ed., Woodhead Publishing Limited, 2017, p. 109.

Cauvain, S.P. "Cakes, Sponges and Muffins." Baking Problems Solved, 2nd ed., Woodhead Publishing Limited, 2017, p. 277.

Edwards, W.P. "Science." The Science of Bakery Products, The Royal Society of Chemistry Publishing, 2007, p. 13.

Timberlake, K. "Acids and Bases." Basic Chemistry, 5th ed., Pearson Education Limited, 2017, pp. 490–496.

Water Absorption Power (WAP)

Water Absorption

Also Known as Water Hydration or Water Absorption Capacity (WAC)

What is Water Absorption?

Water absorption is the amount of water taken up by flour to achieve the desired consistency and create a quality end-product. It is the optimal amount of water you can add to a dough before it becomes too sticky to process.

Water absorption is usually defined by flour weight. For example, 60% water absorption means 60 kg of water is used to hydrate 100 kg of flour.

water absorption works when flour and water are mixed together, water molecules hydrate the gluten-forming proteins gliadin and glutenin, as well as damaged starch and the other ingredients. The hydration process is achieved when protein and starch molecules create hydrogen bonds and hydrophilic interactions with the water molecules.

Particles hydrate by rubbing against each other and contacting water. Process parameters such as type of mixer, beating arm, water flow and pressure remove the hydrated surface layer and expose a new layer of the particle to the excess water so the water diffusion process can continue.

Function

Water absorption influences the following parameters of baking and bread:

- Loaf volume1
- Fracture stress of bread crumb1
- Bread yield2
- Machinability3
- Proofing3
- Final products attributes3
- Shelf life3

Measuring Water Absorption

A farinograph is usually used and can provide an approximate value for water absorption.

The test is based first on the determination of the optimal amount of water to achieve a consistent Brabender Unit value or BU. The amount of water required to produce this consistency is the water absorption of a flour. The analysis of the typical farinograph curve yields important quality parameters such as:

- Arrival time to first achieve 500 BU
- Optimum mix time, also called dough development time
- Departure time
- Mixing Tolerance Index
- Dough stability4

Application

Water absorption levels will vary from 50–54% in a cookie formula, 60–62% in a standard white bread formula, and 80-90% in an artisan-type Ciabatta formula. When developing with sprouted flour, the increase in enzymatic activity in the flour will increase water absorption and maillard reaction.

Influencing factors of flour water absorption:

- **Starch**: About 46% of the total water absorbed is associated with the starch.
- **Proteins**: About 31% of the total water absorbed is associated with the proteins.
- **Pentosans**: About 23% of the total water absorbed is associated with the pentosans.
- **Vital wheat gluten (VWG)**: When added to the formula, it will increase the water absorption and impart greater stability to the dough.5
- Other water binding ingredients like fiber, eggs, bran and hydrocolloids (gums).

Under- and over-absorption in bakery products

Different bakery products require different water absorption or hydration levels. Each level is unique for each kind of baked good to achieve the desired dough consistency and finished product characteristics.

The following are common problems that can occur during mixing. Both under- and over-absorption lead to quality issues in the dough and finished product:6

Under-absorption

Dough	Finished product
• Stiff and dry • Not enough water to hydrate and disperse dry ingredients • Final development cannot occur • Proofs slower • Inadequate pan flow	• Low volume • Dense and firm internal structure (crumb) • Stales more quickly • White corner edges

Over-absorption

Dough	Finished product
• Wet and sticky • Extended clean-up time • Prone to over-fermentation • Diminished tolerance to handling (i.e., poor machinability)	• Poor symmetry • Large volume • Open crumb grain with large cells • Prone to mold

References

Zghal, M. C., M. G. Scanlon, and H. D. Sapirstein. "Effects of Flour Strength, Baking Absorption, and Processing Conditions on the Structure and Mechanical Properties of Bread Crumb." Cereal Chemistry Journal 78.1 (2001): 1-7.

Puhr, D. H., and B. L. D'Appolonia. "Effect of Baking Absorption on Bread Yield, Crumb Moisture, and Crumb Water Activity." Cereal Chem. 69.5 (1992): 582-86.

Pyler, E. J. "Physical and Chemical Test Methods." Baking Science & Technology. Merriam, Kan.: Sosland Pub., 1988. 851-71.

Serna-Saldivar, S.O. "Quality Control of Cereal Grains and Their Products." Cereal Grains: Properties, Processing, and Nutritional Attributes, CRC Press, Taylor & Francis Group, 2010, pp. 491–493.

Sumnu, S.G., and Sahin, S. "Functions of Ingredients in the Baking of Sweet Goods." Food Engineering Aspects of Baking Sweet Goods, CRC Press, Taylor & Francis Group, 2008, pp. 37–39.

AIB International. Bread Manufacturing Process, Distance Learning Course, "Module 02 – Mixing."https://www.aibonline.org/Start-Your-Training/Baking/Baking-Foundations/Bread-Manufacturing-Process-Online, 2018.

BREAD IMPROVER

Bread improver is a blend of ingredients that activate the gluten and help produce gas, which assists and improves the processes of dough kneading and fermentation. The result is a lighter loaf with better texture and keeping qualities. They are used more often in grain mixes or breads with addition of fruit, seeds or nuts to a loaf to give strength and volume. Lacked include bread improver in our bread mixes, but it can also be purchased separately.

Bread improver are substances, which when added to dough, enables the baker to produce an improved loaf with better keeping qualities, finer textures, softer crumb, added bloom and enhanced flavor.

There are three main types of bread improvers:

1. Mineral additives
2. Yeast foods
3. Enriching agents

1.Mineral Additives

Mineral bread improvers are used during the milling of wheat flour. The baker commonly uses them during production as well. They include:

Perusulphates

Used by the miller at the rate of ¼ or ½ oz. per 280 lbs. (one sack) the perusulphates used are potassium and ammonium. Flour treated with persusulphates will take on more water and an increased yield is obtained.

Glycerol Mono Stearate

The mono glycerol ester of stearic acid, which has remarkable emulsifying power, is used as an emulsion stabilizer and as a crumb softener in bread.

Potassium Bromate

It is used by the miller at the rate of 1 lb. per sack (280 lbs.). Bromate increases the stability on the gluten to extend. Bromate has an astringent action on gluten thereby increase the use of water in the dough. It also increases the gas retaining properties of the gluten, thus improving loaf volume.

Phosphates

Acid calcium phosphates and ammonium phosphates both have a tightening action on gluten and since phosphates are a necessary constituent of yeast food, they are both fermented stimulants. Acid calcium phosphate (ACP) is used at the rate of lb. per sack (280lbs) which can be increased to 2 lbs. per sack to inhibit the development of ROPE. A phosphate is added at the rate of 8 oz. per pack.

Lime Water

Limewater was used to retard the fermentation of the dough in hot weather climates. In addition, it has astringent action on the gluten. As lime is alkaline, it reduces the acidity of the dough and thus and shows the rate of fermentation, it is used at the rate of 1 quart per sack.

Organic Acid

Organic acids are natural constituents of fermented dough. They are added to get the dough better conditioned. Lactic acid can be added to the rate of 8 oz. per sack. Sussinic acid is added at the rate of 2-4 oz. per sack.

2. Yeast Foods

Yeast foods indirectly affect the bread in a number of ways by their effect on fermentation. Malt not only provides food directly to the yeast but also manufactures further supplies as and when needed whilst simultaneously mellowing and softening the gluten of the flour.

There are two types of malt: **diastatic** and **non-diastatic**.

Diastatic Malt add to the flavor, it increases the sugar content in the dough and provide diastatic sugar for the fermentation process. Diastatic enzymes also contain proolytic enzymes, which modify gluten.

Non-Diastatic Malt serves the dual purpose of providing sugar as well as during to the flavor.

Flour contains natural sugar. Principally, this is sucrose in varying amounts. Normally, it is 2.5-3%. This amount is not sufficient for satisfactory fermentation. There must be sufficient sugar present for the production of gas that will give the loaf the required volume and to allow for the caramelization of the crust during the baking. As sugar contains no nitrogen, they cannot be considered complete foods for yeast, but they produce material from which CO2 can be produced. Demerara sugar can be used in brown breads, as they are excellent for importing flavor and retaining color.

3. Enriching Agents

Enrichment is a way of increasing nutritional value of the bread along with improvements in the volume. Texture and the keeping quality of the bread.

Fats

Fats have a physical rather than a chemical effect on dough. As fat is a shortening agent, it reduces toughness, thus making the product more mellow. It is particularly valuable for use with strong flour with a tough and harsh gluten content. Fats can be used in small quantities to give optimum effect. Fat also increase food value. They add to the moistness in the bread thereby retarding staling. They also impart flavor to the bread.

Milk and Milk Products

Whole milk added to dough has the effect of adding fat as well as sugar, besides calcium salts and casein.

Eggs

The incorporation of eggs in a bread dough results in many improvements. Egg adds to the increased volume better texture and better oven spring. It is economical to use as it contributes immensely to improved quality and volume of the product.

The term 'bread improver' is used to embrace a wide range of materials that can be added to wheat flour and dough in order to improve some aspect of dough behavior and final bread quality. The use of the term is common and most often applied to the addition of several ingredients at low levels blended with a 'carrier', a material that may or may not have functional properties but that aids dispersion and provides a more conveniently handled composite material. The formulation of bread improvers will be influenced by legislative control over the list of permitted ingredients that may be used in bread making.

Alternative names for bread improvers that may be encountered in the baking industry include:

- Dough conditioners, a specific reference to the fact that the material addition changes dough rheology.
- Processing aids, that implies a similar function to dough conditioners.
- Oxidising agents, that implies a more specific role concerned with the formation of the gluten network in the dough.
- Additives, more commonly applied to specific ingredients.
- Concentrates, similar to an improver but with a greater range of ingredients present (e.g. fat, sugar and salt) and commonly used at higher rates of addition.

Almost any material added to a flour and water dough will have some improving effect. For example, the addition of yeast improves the lightness and palatability of bread, while salt changes the handling properties of wheat flour doughs and the flavor of the baked bread. However, the term bread improver is now commonly restricted to materials that are typically added at much lower levels of addition than yeast or salt with the intention of improving gas production or gas retention in the dough, retaining bread crumb softness and obtaining a whiter crumb color.

Some of the more common ingredients used in bread improvers are noted below. The classification used is arbitrary since the complex actions of most materials in bread making means that they might be classified in more than one group. For example, the addition of many enzyme preparations brings about changes in dough rheology that makes it easier to process doughs but also results in improved oven spring, a manifestation of improved gas retention.

- Aids to dough processing: enzyme-active preparations, e.g. malt flour, fungal a-amylase.
- Aids to gas production: yeast foods, such as ammonium chloride.
- Aids to gas retention: oxidizing agents, such as ascorbic acid and potassium bromate.
- Aids to bread softness: e.g. glycerol monostearate (GMS).
- Aids to improving crumb colour: soya (soy) flour.

SUGAR

Sugar and Sweeteners

Granulated Sugar

This Sugar has been refined into crystals from either sugarcane or sugar beets, superfine granulated sugar is preferred in baking because of its ability to dissolve faster in cake batters and other mixtures. It is often used for making meringues.

Powdered Sugar

This sugar also known as confectioner's sugar. It has a very fine texture, the same as flour, and must always be sifted before adding to other ingredients. It should always be stored in airtight container.

Raw Sugar

Raw sugar is less refined then other sugars, having a coarse lumpy texture. It has not been bleached and so retains its light brown to blond colour.

Brown Sugar

Brown sugar available in light and dark colours. The higher the percentage of molasses and other solids, the darker in colour and more intense flavour. It should be store in airtight container. To keep brown sugar soft, place a slice apple or potato in an open plastic bag inside the container with the brown sugar.

Granulated Brown Sugar

Granulated brown sugar is not compact like brown sugar. Moisture is removed during the processing and refining, allowing the granulation. It is not possible to replace brown sugar with granulated brown sugar, because of the difference in moisture contain. The resulting product will have a different texture. It is used most often as a topping for pancakes and fruits.

Molasses

This a thick brownish liquid by product derived from the refining of granulated sugar. The lighter the colour , the sweeter the molasses.

Maple Sugar

Maple sugar is refined male sap and has a rich, wonderful flavour. It is highly prized product which is very expensive. Maple sugar is a great topping for Danish pastries and special desserts and cakes.

Sugar Substitutes

In addition to adding sweetness and flavour, sugars perform the following functions in baked goods:

a) They create tenderness and fineness of texture.

b) They retain moisture, thus improving texture and keeping qualities.

c) They give crust colour because of their browning properties.

When sugar in a formula is reduced, these functions must be performed by other ingredients.

Some customers who want to reduce or eliminate **refined white sugar** in their diets have no objection to **natural sugars like honey**, even though they may have the same number of calories.

Another **granular sugar**, such as **date sugar or maple sugar**, can be substituted for white sugar in equal weights without other changes to the formula.

If a liquid sugar is used, such as honey, brown rice syrup, molasses, or maple syrup, the following adjustments to the formula may be necessary:

a) Liquid sugars have no creaming ability, so other forms of leavening must be substituted. You may be able to mix the batter by the muffin method rather than the creaming method, as long as you increase the quantity of baking powder.

b) If a large quantity of liquid sugar is used, reduce the other liquids in the formula.

c) Not all liquid sugars have the same sweetening power, so you may have to adjust quantities. Brown rice syrup, for example, is only 30 to 60 percent as sweet as white sugar, whereas honey is sweeter than white sugar.

Sugar substitutes are used when it is necessary to reduce the total quantity of sugar in a formula.

Sugar substitutes are chemicals that have a sweet taste, usually much sweeter than sugar, but no or little nutritional value.

Aspartame and saccharin are intensely sweet chemicals often used as beverage sweeteners, but they have little use in baking because they don't perform any of the functions of sugar listed above, other than providing sweetness. Also, aspartame is destroyed by heat, so it is of no use in baked goods. Furthermore, these sweeteners leave aftertastes that are unpleasant to many people, and there are also questions about their safety.

Coconut Sugar

Coconut palm sugar has a lower glycaemic index (GI) than regular sugar, it comes from coconut tree sap.

People with diabetes may prefer it to regular sugar for this reason, but the carbohydrate and calorie counts of both are the same. Try sprinkling some coconut palm sugar on your oatmeal or popcorn for added sweetness.

Fruit Concentrates

Fruit juice concentrates, such as apple juice concentrate, orange juice concentrate, or white grape juice concentrate, are wonderful substitutes for sugar and add interesting flavors as well.

Juice concentrates are made up of fructose and glucose.

Use ¾ cup for every cup of white sugar, and decrease the amount of liquid by 3 tablespoons.

Date Sugar or Maple Sugar

Date sugar is a type of sugar found most commonly in natural food stores since it is less processed than more conventional sugars, date sugar is a less refined sugar than typical white sugar.

It is made from dried dates and adds a rich sweetness to recipes, although it will not dissolve when added to drinks.

It also does not melt like granulated sugar which can limit its use.

It is sometimes promoted as a healthier alternative to brown sugar, although it can be quite expensive.

Date sugar should not to be confused with date palm sugar or also called palm sugar as this is made from the sap of the sugar palm tree, including date trees.

Date sugar is made by first making a paste from raw dates. Then, the paste is mixed with a substance called maltodextrin (a common food additive). This mixture is oven dried and ground into granules.

Date sugar can be substituted in many foods and beverages.

Refined Fructose

Refined fructose is sweeter than granulated sugar. It can be easily substituted in baking recipes--simply add one-third less.

Some tasters find that, although products made with fructose taste sweet, they also taste a little flat.

Fructose attracts more water than sucrose, so fructose-sweetened products tend to be moist.

Baked products made with fructose will be darker than if they were made with white sugar. Fructose is available in health-food stores.

Isomalt

Isomalt is another sugar substitute. It is white and granular in appearance and can be substituted for an equal weight of regular sugar.

Isomalt has only half the calories of sugar, but it also has only half the sweetness, so it is not a suitable substitute in all formulas. In addition, it is not easily digested and can cause intestinal discomfort and bloating when eaten by some people.

Isomalt basically used in baking for only sugar work like sugar show pieces.

Artificial Sweeteners

These sweeteners have been approved by the FDA and are available for home use.

While they provide a sweet taste, artificial sweeteners lack the browning, tenderizing and moisture-retaining properties of granulated sugar.

Sucralose is the one sweetener than can be substituted cup-for-cup for granulated sugar in baking.

1. Sucralose

Sucralose is the most useful sugar substitute in baked goods. It is sold under the brand name Splenda.

Pure sucralose is 600 times sweeter than sugar.

For baking, it is mixed with a bulking agent called maltodextrin to give it the same sweetening power and texture as an equal volume of sugar. This product is called granular sucralose.

In pies, cookies, quick breads, dessert sauces, and custards, substitute an equal volume of granular sucralose for the sugar in the formula. (Granular sucralose has 96 calories per cup/240 mL, primarily from the bulking agents, while granulated sugar has 770 calories for the same volume.)

Granular sucralose, however, does not have good creaming abilities, it does not contribute to browning, it does not contribute to texture, and it does not improve keeping quality as sugar does.

When these functions are important, the usual technique is to substitute granular sucralose for half of the sugar in the formula, thus reducing the number of calories from sugar by half.

A so-called baker's blend, consisting of half-granular sucralose mixed with half sugar, is available, but it is more cost-effective to make the blend yourself.

Please note that when you are substituting granular sucralose for sugar, you must substitute an equal volume, not an equal weight, because the granular sucralose is much lighter than sugar.

One cup (240 mL) granular sucralose weighs about 25 grams, slightly less than 1 ounce.

When baking with sucralose, monitor the products carefully for doneness. You can't rely on the usual amount of crust browning as a doneness indicator because the product won't brown as well.

Recipes made with this product tend to bake faster than usual, so check for doneness sooner than the recipe specifies.

2. Saccharine

It is 200 to 700 times sweeter than sugar. It can be used in baked goods.

However, the manufacturer recommends substituting it for only half of the sugar in a recipe. Substitute 6 (1-gram) packets for each ¼ cup sugar.

It is sold under the brand name Sweet and Low®.

3. Aspartame

It is 160 to 220 times sweeter than granulated sugar.

This sweetener is heat-sensitive: it loses its sweetening power when heated, and cannot be used for cookies or cakes.

The manufacturer does recommend trying it in no-bake pies and in puddings after they have been removed from the heat.

Substitute 6 (1-gram) packets for each ¼ cup of sugar.

It is sold under the brand names Equal® and NutraSweet®.

4. Acesulfame Potassium

It is 200 times sweeter than sugar.

It is heat-stable, so it can be used in baking and cooking. Use acesulfame K in combination with granulated sugar when baking. Substitute 6 (1-gram) packets for each ¼ cup sugar.

It is sold under the brand names Sunette® and Sweet One®.

Syrups

Honey

Honey is a natural sugar syrup consisting largely of the simple sugar's glucose and fructose, plus other compounds that give it its flavour and colour.

Honey is 25 to 50% sweeter than sugar, varies considerably in flavour & colour, depending on their source. Flavour is the major reason for using honey, especially as it can be expensive. Because honey contains invert sugar, it helps retain moisture in baked goods. Like molasses, it contains acid, which means it can be used with baking soda as a leavening.

Baked goods made with honey are moist and dense, and tend to brown faster than those made with granulated sugar. Use ¾ cup plus 1-tablespoon honey in place of 1-cup sugar, and reduce the other liquid ingredients by 2 tablespoons. Unless the recipe includes sour cream or buttermilk, add a pinch of baking soda to neutralize the acidity.

Molasses

Molasses is concentrated sugarcane juice. Sulphured molasses is a by-product of sugar refining. It is the product that remains after most of the sugar is extracted from cane juice. It contains small amounts of B vitamins, calcium, and iron.

Unsulfured molasses is not a by-product but a specially manufactured sugar product. It has a less bitter taste than sulphured molasses.

Molasses contains large amounts of sucrose and other sugars, including invert sugar. It also contains acids, moisture, and other constituents that give it its flavour and colour.

Darker grades are stronger in flavour and contain less sugar than lighter grades. Molasses retains moisture in baked goods and therefore prolongs freshness.

Crisp cookies made with molasses can soften quickly because the invert sugars absorb moisture from the air.

When substituting molasses for sugar, use 1 1/3 cups molasses for 1-cup sugar, and reduce the amount of liquid in the recipe by 5 tablespoons.

Molasses is also more acidic than sugar; add ½-teaspoon baking soda for each cup of molasses used. Replace no more than half the sugar called for in a recipe with molasses.

Brown Rice Syrup

Brown rice malt syrup consists of maltose, glucose and complex carbohydrates. It is an amber-hued syrup resembling honey, but it is not as sweet as honey.

It can be substituted cup per cup for granulated sugar, but the liquid ingredients should be reduced by ¼ cup per cup of rice syrup. Enzyme-treated syrup, as opposed to malted syrup, will tend to liquefy the batter of a baked product. Use the malted syrup for best results.

Corn Syrup

Corn syrup is known as an invert sugar; it is useful in cooking and candy making because, unlike other sugars, it does not crystallize. Corn syrup is less sweet than sugar, and does not add flavour like molasses or honey.

Golden Syrup, common in the United Kingdom, is a refinery syrup made from sugar. It is used in place of corn syrup.

Sugar syrups have a livelier flavour than corn syrups and add more character to dishes such as pecan pie.

Maple Syrup

Maple syrup is made from the sap of sugar maple trees. The sap is boiled down into a sweet, delectable syrup.

Grade A maple syrup is golden brown and has a delicate flavour. Grade B is thicker, darker, and is better for baking because it has a stronger flavour--and it costs less.

Maple syrup contains a fair bit of sugar. It has nutrients such as antioxidants, calcium, iron and potassium, and you can cut your sugar consumption by about 33% by using maple syrup instead.

Maple syrup is also friendlier toward your blood sugar, with a glycaemic index of 54 versus table sugar's 65.

If you make your own dairy-free milk, try sweetening it with a touch of maple syrup.

Although maple syrup is only 60% as sweet as sugar, use ¾ cup for every cup of white sugar and decrease the amount of liquid by 3 tablespoons to compensate for its liquid state.

Agave Nectar or Agave Syrup

Agave syrup, aka maguey syrup, and commonly though inaccurately known as agave nectar, is a sweetener commercially produced from several species of agave, including Agave tequilana (blue agave) and Agave salmiana.

Blue-agave syrup contains 56% fructose as a sugar providing sweetening properties. It is 1.4 to 1.6 times sweeter than sugar, and may be substituted for sugar in recipes, because it comes from a plant, it is widely utilized as an alternative to honey for those following a vegan lifestyle, and is often added to some breakfast cereals as a binding agent.

The specific strain known as blue agave syrup is not recommended for people with fructose intolerance.

Agave syrups are sold in light, amber, dark, and raw varieties.

Light agave syrup has a mild and almost neutral flavour, and is therefore sometimes used in delicate-tasting dishes and beverages.

Amber agave syrup has a medium-intensity caramel flavour, and is used in dishes and drinks with stronger flavours.

Dark agave syrup has even stronger caramel notes and imparts a distinct flavour to dishes, being used in some desserts, poultry, meat, and seafood dishes.

Both amber and dark agave syrups are sometimes used "straight out of the bottle" as a topping for pancakes, waffles, and French toast.

The dark version is unfiltered and therefore contains a higher concentration of the agave plant's minerals.

Jams

Jams made by boiling fruit and sugar to a thick consistency.

Jams are made from the entire fruit, including the pulp, while preserves are essentially jellies that contain whole or large pieces. Jams are made from crushed or ground fruit and usually have a thick consistency due to high pectin content.

Jam is a condiment. Most jams are cooked. After making, jam is normally put into an airtight jar. Usually a jam contains as much sugar as it contains fruit. The two parts are then cooked together to form a gel.

As a gel, jam is neither a solid or a liquid. It can contain chunks of fruit which are solids. Once opened and out of the fridge the gel becomes more like a liquid, and is able to decay.

As per FSSAI, Jam means the product prepared from sound, ripe, fresh, dehydrated, frozen or previously packed fruits including fruit juices, fruit pulp, fruit juice concentrate or dry fruit by boiling its pieces or pulp or puree with nutritive sweeteners namely sugar, dextrose, invert sugar or liquid glucose to a suitable consistency. It may also contain fruit pieces and any other ingredients suitable to the products. It may be prepared from any of the suitable fruits, singly or in combination. It shall have the flavour of the original fruit(s) and shall be free from burnt or objectionable flavours and crystallization. The specific requirements for Jam are as follows:

Total soluble solids (m/m) - Min. 65.0 %

The product shall be manufactured from minimum 45.0% by weight, of original prepared, fruit, exclusive of any added sugar or optional ingredients of finished product, except for Strawberry or Raspberry fruit, where it shall contain minimum 25.0% fruit.

Salient Characteristics of Jam

The amount of pectin in the gel is quite small (< 1.0% of the weight of jam); sugar content is very high (60-70%). Of the remainder, the bulk is water with a small amount of fibrous matter and seeds.

For a chunky texture, some pieces of fruit are added about 5 minutes before the cooking is finished. If a stiffer or sweeter jam is desired, more thickener or pectin, honey or sugar may be added.

Fruit or Fruit Combinations

Some fruits such as lemons and bitter oranges are rich in both acid and pectin and can be easily made into jam. Some popular combinations include: pineapple and paw paw, orange and banana, apple and plum.

Sugar

55 parts of sugar is required for every 45 parts of fruit. With excess sugar, the jam becomes gummy and sticky. The finished jam should have 30-50% invert sugar / glucose to avoid crystallization of cane sugar during storage.

Acid

Acidity can be supplemented by use of citric, tartaric or malic acid. The pH influences inversion of sugar and setting of the jam. For pectin jam, pH of 3.3 gives a good set. For Apple and Plum jam, the pH range of 3.2-3.5 is desirable: optimum being 3.35.

Preparation of Jam

Fresh Ripped Fruits

↓

Inspection

↓

Washing

↓

Peeling and Cutting

↓

Blending

↓

Simmering

↓

Stirring (To draw out Pectin in Fruits)

↓

Addition of Sugar/Water

↓

Rapid Boiling/Concentration + Lime

↓

Jam Product

↓

Filling and Packaging into Container

↓

Cooling for Storage in a dry Place

The process for preparing Fruit Jam is depicted in Flow chart

Storage of Jam

The surface of jam is susceptible to mold growth; yeast cannot grow or thrive. There is risk of mold development and fermentation, with alteration in the taste. Permitted preservatives may be used at levels sufficient to inhibit the growth of yeast and molds.

Jam should be stored in a fairly cool place or else some moisture evaporates resulting in surface graining and shrinkage of jam. It should be stored in a place having RH of ~ 80%.

Marmalades

The word marmalade is borrowed from the Portuguese marmelada, from marmelo quince. Marmalade, usually made from citrus fruit, is a jellylike concentrate of prepared juice and sliced peel. It is a jelly with pieces of fruit suspended in it. Citrus peel and juice are frequently the basis of marmalade. The well-known version is made from bitter orange, but it is also made from lemons, limes, grapefruits, mandarins, sweet oranges, bergamots and other citrus fruits, or a combination.

As per FSSAI, Marmalade means a product prepared by boiling sound fruits with peel, pulp and juice, with or without water, added nutritive sweeteners and concentrating to such a consistency that gelatinization takes place on cooling of the product. It shall not be syrupy, sticky or gummy and shall be clear and transparent.

The specific requirements are as follows:

Total soluble solids (m/m) - Min. 65.0 %

Fruit content except peel (m/m) - Min. 45.0 %

Peel in suspension - Min. 5.0 %

The container shall be well filled with the product and shall occupy not less than 90% of the net weight of the container, when packed in the rigid containers.

Marmalade is generally distinguished from jam by its inclusion of fruit peel.

Cocoa Production

Chocolate is produced from the seeds of a tropical tree called the cocoa or, more properly, cacao tree.

Cocoa trees produce large pods full of seeds called cocoa beans.

After the pods are harvested, the beans are quickly removed and allowed to ferment until they lose most of their moisture. There are several ways of doing this, but the traditional method is to spread them between layers of banana leaves and leave them for several days, turning them often so they ferment evenly.

The chemical changes that take place during fermentation turn the beans from yellowish to brown and begin to develop the flavour.

The fermented beans are next dried in the open air, because they still contain a great deal of moisture.

The dried beans are then ready to be shipped to processors. A single tree yields only 1 to 2 pounds (500 g to 1 kg) dried beans.

Cocoa processors clean the dried beans thoroughly and then roast them. The true flavour of the cocoa develops during roasting, making the temperature and degree of roasting important factors in the quality of the finished chocolate.

After roasting, the beans are cracked and the shells are removed. The broken particles of cocoa that result is called nibs. Nibs contain more than 50 percent fat, in the form of cocoa butter, and very little moisture.

Grinding the nibs produces a paste and releases the cocoa butter from inside the cell walls. This paste is called chocolate liquor or cocoa mass and is the basis of chocolate production.

When chocolate liquor cools, it sets into a hard block.

The next stage of manufacturing is to separate the cocoa powder from the cocoa butter. This is done with powerful hydraulic presses that squeeze out the melted fat, leaving hard cakes that are then ground into cocoa powder. Meanwhile, the cocoa butter is purified to remove odor and colour.

Cocoa By-Products

Cocoa Powder

Cocoa is the dry powder that remains after part of the cocoa butter has been removed from chocolate liquor.

Dutch process cocoa, or dutched cocoa, is processed with an alkali. It is slightly darker, smoother in flavour, and more easily dissolved in liquids than natural cocoa.

Dutched cocoa, on the other hand, is generally neutral or even slightly alkaline. Therefore, it does not react with baking soda (see the Baking Soda Needed to Balance the Acidity of Typical Cocoa Products table). Instead, baking powder is used as the sole leavening agent.

Natural cocoa is somewhat acidic. When it is used in such products as cakes, it is possible to use baking soda (which reacts with acid) as part of the leavening power.

If you are substituting dutched for natural cocoa, you must increase the baking powder by 1 ounce (30 g) for each 1/2 ounce (15 g) soda omitted.

If not enough soda is used in chocolate products, the colour of the finished product may range from light tan to dark brown, depending on the quantity used.

If too much is used, the colour will be reddish brown. This colour is desired in devil's food cakes but it may not be wanted in other products. When switching from one kind of cocoa to another, you may have to adjust the soda in your recipes.

Cocoa Butter

Cocoa butter is the fat pressed out of chocolate liquor when cocoa is processed. Its main use in the bakeshop is to thin melted couverture to a proper consistency.

Not all the cocoa butter is pressed out. Therefore, cocoa powder contains some fat in the form of cocoa butter.

Regular cocoa has a fat content of 10 to 12% and is known as 10/12 cocoa.

High-fat cocoa has a fat content of 22 to 24% and is known as 22/24 cocoa.

Both types are used in the bakeshop. For hot cocoa beverage, 22/24 cocoa is more often used.

Low-fat cocoa, with a fat content below 10%, must be specially labelled and is more difficult to manufacture. It is expensive and not generally used in the bakeshop.

EGGS AND DAIRY PRODUCTS

Eggs

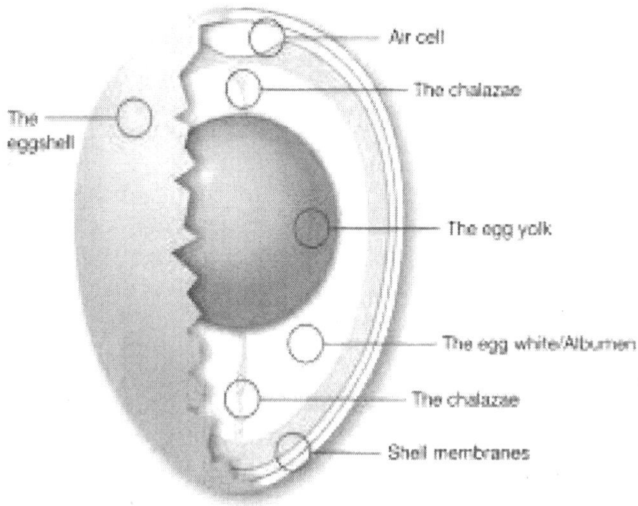

Eggs should be well understood by the baker because they are used in large quantities in the bakeshop and are more expensive than many of the other high-volume ingredients, such as flour and sugar.

For example, half or more of the ingredient cost of the average cake batter is for the eggs.

Composition

A whole egg consists primarily of a yolk, a white, and a shell. In addition, it contains a membrane that lines the shell and forms an air cell at the large end, and two white strands called chalazae that hold the yolk centered.

The yolk is high in both fat and protein and contains iron and several vitamins. Its colour ranges from light to dark yellow, depending on the diet of the chicken.

The white is primarily albumin protein, which is clear and soluble when raw but white and firm when coagulated. The white also contains sulphur.

The Average Composition of Fresh Liquid Eggs table lists the average water, protein, and fat content of whole eggs, whites, and yolks.

Types of Egg

All eggs sold at retail in Canada are antibiotic and hormone free. All eggs are produced to the same world-class Canadian standards for food safety and quality. All Canadian eggs are fresh, nutritious and delicious!

- **Standard White Eggs**

  These eggs come from white hens that are typically raised in conventional housing systems. Conventional housing has been the standard for decades, but the Alberta egg industry has begun to transition away from conventional housing systems.

- **Standard Brown Eggs**

  These eggs come from brown hens that are typically raised in conventional housing systems. Conventional housing has been the standard for decades, but the Alberta egg industry has begun to transition away from conventional housing systems.

- **Furnished / Enriched / Nest-Laid Eggs**

  These eggs come from hens that are raised in furnished housing systems. Furnished housing provides more space (both floor space and height) for the hens to move around, while also providing a variety of enrichments, which allow the hens to express more natural behaviors. Enrichments include nesting boxes, perches, scratch pads and dust baths.

- **Free-Run Eggs**

  These eggs come from hens that are raised in free-run (barn or aviary) housing systems. Free-run systems allow the hens to roam freely within an enclosed barn, while also providing a variety of enrichments such as nesting boxes and perches.

- **Free-Range Eggs**

  These eggs come from hens that are raised in free-run (barn or aviary) housing systems, which also provide access to outdoor runs (when weather permits).

- **Organic Eggs**

  These eggs come from hens that are raised in free-range housing systems. The hens are only provided feed that has been certified organic, which means that it only contains ingredients that were grown without pesticides, herbicides or commercial fertilizer. Look for a certified organic symbol on the egg carton.

- **Omega-3 Eggs**

  These eggs come from hens that were provided feed containing extra flax (up to 10-20%). As a result, the eggs laid by these hens contain more Omega-3 fatty acids.

- **Vitamin-Enhanced Eggs**

  These eggs come from hens that were provided feed that was nutritionally enhanced to include higher levels of certain vitamins (i.e.: Vitamin D or Vitamin E). As a result, the eggs laid by these hens contain corresponding higher amounts of the particular vitamin(s).

- **Vegetarian Eggs**

  These eggs come from hens that were provided feed containing only plant-based ingredients.

- **Processed Eggs**

  Liquid, frozen and dried egg products come from eggs that were broken by special egg breaking machines, and then pasteurized. They often contain added ingredients (i.e.: preservatives, flavoring or coloring) and this process allows for special formulations (i.e.: egg whites only).

Importance in Bakery

Eggs play an **important role** in everything from **cakes** and cookies to meringues and **pastry** cream — they create structure and stability within a batter, they help thicken and emulsify sauces and custards, they add moisture to **cakes** and other **baked goods**, and can even act as glue or glaze.

Grades and Quality Grades

In the United States, eggs are graded for quality by the USDA. There are three grades: AA, A, and B.

The best grade (AA) has a firm white and yolk that stand up high when broken onto a flat surface and do not spread over a large area. As eggs age, they become thinner and are graded lower.

In Canada, there are four egg grades: A, B, C, and Canada Nest Run. As a baker, you will not be concerned so much with the firmness of yolks and whites.

Rather, you will want eggs that are clean and fresh-tasting, free of bad odors and tastes caused by spoilage or absorption of foreign odors. One bad-smelling egg can ruin an entire batch of cakes.

Maintaining Quality

Proper storage is essential for maintaining egg quality.

Eggs keep for weeks when held at 36°F (2°C) but lose quality quickly when held at room temperature.

In fact, they can lose a full grade in one day at warm bakeshop temperatures. There is no point in paying for Grade AA eggs if they are Grade B by the time you use them.

Store eggs away from other foods that might pass on undesirable flavors or odors.

Egg Selection

Selection of good eggs for hatching is important to achieve good results. First priority should be given to eggs from hens which are between eight and eighteen months of age. This is because the size, interior and shell quality of eggs from this age group is most likely to be good for hatching. Next, look at the shell closely. Eliminate any eggs which are cracked, dirty or misshapen.

Dairy Products

Whole Milk

This milk that contains 3.5% butter fat and is used in bakingfor the rich flavour which provides.

Low Fat Milk

Low fat milk contains 1% to 2% butterfat. When substituted for whole milk, it may be necessary to consider the change in fat content.

Skim Milk

This milk contains 0.1% to 0.5% butterfat, and is not suitable replacement for whole milk.

Buttermilk

This is the product remaining after sweet or sour milk has been churned and the fat removed. It has a distinctive flavour, texture and high acid content.

Cultured Buttermilk

Cultured buttermilk is the soured product produced in the making of pasteurized skim milk after treatment with a suitable lactic acid bacteria culture. This product has a distinctive flavour

Evaporated Milk

This is whole cow's milk from which 60% of water has been removed. It is homogenized, pasteurized and sealed in cans.

Sweetened Condensed Milk

This product is produced by evaporating half of the water from the whole milk and adding sufficient quantities of cane or corn sugar to act as a preservative. It is then heated cooled and canned. This is not a suitable replacement of other milk product due to its high sugar content.

Dry Whole Milk Solids (Milk Powder)

These are the remaining dry solids after all water has been removed from the whole milk.

Yogurt

Yogurt is milk which has been fermented by the introduction of a bectirium.

Heavy Cream

This product contains 35% to 40% butterfat. There is little which replaces the richness of heavy cream.

Light Cream

This product contains less than 30% butterfat.

Guidelines for Using Milk Products in Baking Fresh Liquid Milk

All fresh milk and cheese products must be kept under refrigeration. It should be remembered that liquid milk products are susceptible to odors and flavours and so should be stored in tight containers.

LEAVENING AGENTS

Unilike liquids and solids, gases expand greatly when they are heated. Leavening is theproduction or incorporation of gases in a baked product to increase volume and produce shape and texture. These gases must be retained in the product until the structure is set enough (by the coagulation of gluten and egg proteins and the gelatinization of starches) to hold its shape.

The three main gases that leaven baked goods are carbon dioxide, steam, and air. Two of these gases, steam and air, are present in all baked goods. An essential part of the leavening process is the formation of air cells during mixing. Even if a properly mixed dough or batter appears dense and compact, it actually contains millions of tiny air cells. Leavening gases are trapped in these air cells, which expand as they fill with gases and as the gases are heated. The walls of the cells are formed largely from gluten proteins and sometimes egg proteins. They hold the gases and form the structure of the baked item.

Yeast

What Is Yeast?

Yeast is a microorganism that is grown in yeast factories. Its scientific name is Saccharomyces cerevisiae. In this word you can see the Latin word "saccharo", which means sweet or sugar and the word "myces", which means, "mold". Baker's yeast is a unicellular mold that reproduces through a process, which is known as "budding".

We know that bakers use it to make the dough "rise"; without it, our bread would be like flat, hard cakes. In the days when people made their own bread, they would go to a brewer and get a jug of brewer's yeast. It was fluid and yellow. Nowadays, yeast is made commercially on a large scale. The yeast you buy at your market, the yellow lumps done up in paper, has been compressed for convenient handling.

Commercial yeast obviously is widely available. The baking characteristics of the yeast are the most significant feature of commercial yeast. Different types of yeast are available for different types of bread-making processes (f.i. osmotolerant yeast suitable for doughs that contain a lot of sugar). Properties such as the shelf life of the yeast are controlled by various fermentation parameters. For instance, a high percentage of budding cells at the time of the harvest will result in a product with a shorter shelf life.

Yeast is a plant, according to the biologists, and is capable of reproducing itself. A piece of yeast consists of minute cells, with walls composed of cellulose, and an interior of living matter called protoplasm. You can feed it with a solution of sugar to make it grow, or it can be "killed" by starvation or heat. The ancients did not use yeast as we know it today; they prepared a leaven or 'barm' (which has the same action) from ground millet kneaded with "must" out of wine-tubs.

Wheat bran was also used, kneaded with a three-days-old must, dried in the sun, and then made into little cakes. When required for making bread, the cakes were soaked in water, and then boiled with the finest flour, after which the whole was mixed in with the meal. Another old method for making barn was to prepare cakes of barley meal and water; these were baked on a hot hearth, or else in an earthen dish upon hot ashes and left until they turned reddish-brown. Afterwards, the cakes were kept shut up in a vessel until they turned quite sour. When wanted for leaven, they were first steeped in water.

Fermentation

The primary function of yeast is to produce carbon dioxide gas, which expands the dough during proof and the early stages of baking (oven spring). Yeast prefers slightly acid conditions to work best. A pH ranging from 4.5 to 6,0 gives the best results. Bread doughs are generally in the region of pH 5.5 so in normal bread-making the effect of the pH is not a particular consideration. However,some ingredients used in the bakery, such as mold inhibitors in some bread improvers, lower the pH of the dough and do have a retarding effect on yeast fermentation. This effect is usually taken into account when deciding on the yeast level to use in a given recipe. There are other ingredients which can retard yeast activity i.e. spices or raisins.

Carbon dioxide cannot form a gas bubble on its own it requires a "nucleating site" (i.e. somewhere it can gather to form a bubble). In fizzy drinks, microscopic projections on the side of the bottle provide those sites, which is why when you release the pressure as you open the bottle you see "streams" of gas running from the sides. In bread dough the nucleating sites are provided by the nitrogen gas bubbles, trapped in the dough during mixing. The yeast has used up the oxygen from the air.

During proof stages, the carbon dioxide goes into solution until the solution is saturated and then any more, which is generated, makes its way into the nitrogen gas bubbles, which grow in size, and the dough expands. The more yeast and the warmer the temperature the faster the expansion - we get oven spring because the maximum gassing rate occurs at 40 - 45°C.

In bulk fermentation, stages we also get dough expansion from carbon dioxide generation but most of that is lost when the dough is knocked back and divided, so the yeast has to start over again.

Yeast also contributes to dough maturity/development. Though its role is minor compared to improvers in no-time doughs, it is more significant in bulk fermentation where the enzymes, especially the proteolytic ones (they modify the gluten proteins), play a significant role.

To sum up, then, the dough is aerated by the action of the yeast. The little cells we mentioned ferment the dough, and produce tiny bubbles of gas inside it. As a result, the dough gets fatter and bigger, and rises, of course. Thus, when the dough is baked, you have a 'bold' loaf, light and airy; when you cut it, you can see all the tiny holes formed by the gas, so that it looks like a sponge.

Yeasts constitute a group of single-celled (unicellular) fungi, a few species of which are commonly used to leaven bread and ferment alcoholic beverages. Most yeasts belong to the division Ascomycota. A few yeasts, such as Candida albicans, can cause infection in humans. More than one thousand species of yeasts have been described. The most commonly used yeast is Saccharomyces cerevisiae, which was domesticated for wine, bread, and beer production thousands of years ago.

Yeast physiology can be either obligate aerobic or facultative anaerobic. In the presence of oxygen, yeast will multiply through a mechanism known as budding. In the yeast factory this mechanism is used to grow yeast into a commercially available product called compressed yeast. Compressed yeast contains about 75 % water of which about 1/5th sits between the yeast cells and not in the yeast cells. Depending on the moisture content and the size of the individual yeast cells, 1 gram of compressed yeast contains 8 to 13 billion yeast cells. In these circumstances, the yeast does not have access to nutrients and it must survive on the reserves stored in the cell. The yeast cell will slowly but surely use these reserves. As a result, the cells grow weaker and weaker, i.e. the yeast will lose its gassing power. This process is called "autolysis". The speed at which this process progresses, depends on the proteolytic activity of the cell. Proteases are enzymes that split proteins. The proteolytic activity depends on the temperature. The higher the temperature the faster the breakdown of proteins will take place. Therefore, yeast should be stored in the fridge.

In the absence of oxygen, fermentative yeasts produce their energy by converting sugars into carbon dioxide and ethanol (alcohol). In brewing, the ethanol is bottled, while in baking, the carbon dioxide raises the bread, and the ethanol evaporates. There is no known obligate anaerobic yeast.

An example with glucose as the substrate is

$$C_6H_{12}O_6 \text{ -> } 2C_2H_5OH + 2CO_2$$

Yeasts for leavening bread may be produced commercially or caught from the environment. Many types of yeast can be isolated from sugar-rich environmental samples. Some good examples include fruits and berries (such as grapes, apples or peaches), exudates from plants (such as plant saps or cacti). Some yeasts are found in association with insects.

The use of potatoes, water from potato boiling, eggs, or sugar in bread dough accelerates the growth of yeasts. Salt and fats such as butter slow yeast growth down. A common medium used for the cultivation of yeasts is called potato dextrose agar (PDA) or potato dextrose broth. Potato extract is made by autoclaving cut-up potatoes with water for 5 to 10 minutes and then decanting off the broth. Dextrose (glucose) is then added (10 g/L) and the medium is sterilized by autoclaving.

Yeast fermentations comprise the oldest and largest application of microbial technology. They are used for beer and wine fermentations and bread production.

There are many different kinds of yeast, which all belong to the Saccharomyces family. They do not have all the same capacity to ferment sugars. Certain species only ferment one type of sugar, others two different types of sugar. In the case of bread, we need a yeast, which can transform glucose and fructose into CO 2 and alcohol.

It seems appropriate to make a list of all terms used in the bakery and to give a definition of the various expressions.

Yeast Fermentation

Fermentation is the process by which yeast acts on sugars and changes them into carbon dioxide gas and alcohol. This release of gas produces the leavening action in yeast products. The alcohol evaporates completely during and immediately after baking.

Fermentable sugar in bread dough comes from two sources:

1. It is added to the dough.
2. It is produced from flour by enzymes that break down the wheat starch into sugar. These enzymes are present in the flour and/or are added by the baker in the form of diastatic malt, Yeast is a microscopic plant that accomplishes this fermentation process by producing enzymes. Some of these enzymes change complex sugars (sucrose and maltose) into simple sugars. Others change the simple sugars into carbon dioxide gas and alcohol. The following formula describes this reaction in chemical terms:

$$C_6H_{12}O_6 \rightarrow 2CO_2 + 2C_2H_5OH$$

Simple sugar -> carbon dioxide -> alcohol

Types of Yeast

Yeast	Unicellular plant, which metabolizes simple sugars to alcohol and CO_2. This process is called proofing or fermentation. Different species of yeast are used in different types of fermentations i.e. baker's yeast in the bakery, beer yeast in the brewery
Baker's yeast	In 1860, Louis Pasteur discovered why bread rises and what the role of the yeast was in this process. He discovered that a particular type of yeast – Saccharomyces cerevisiae – was particularly apt to be bred industrially.
Natural or wild yeast	Numerous types of yeast can be found on plants, cereals, fruits, vegetables etc. They belong often to the family of Saccharomyces exigus to which belong also the varieties Candida krusei and Candida milleri. Not all natural yeasts are appropriate for use in the food industry.
Lactic acid bacteria	Generic term for a whole series of gram positive bacteria (Enterococcus, Lacto coccus, Leuconostoc, Lactobacillus) which produce mainly or exclusively lactic acid. They are used abundantly in the food industry for the preparation of yoghurt, cheese, sauerkraut, wine, beer, sourdough etc.
Anstellgut (German)	A piece of dough kept from previous productions, which is refreshed at regular intervals by adding fresh flour and water.
Barm	All natural sourdough made from whole-wheat flour or wheat kernels, which begin to ferment spontaneously thanks to the presence of natural lactic acid bacteria and/or natural yeasts.
biga (Italian)	Leaven made on the basis of commercial yeast. Water, flour and a small amount of commercial yeast are mixed. The consistency is the same as normal dough. The mixture is allowed to ferment for 24 hours and subsequently added to freshly made dough, to which, if necessary, more yeast is added.
Chef	A piece of dough kept from previous productions. It is used to make "levain". A "chef" however is normally less solid than levain
levain (French)	Is made from the "chef" to which more flour and water is added in order to get the consistency of normal dough. Normally it is kept at relatively low temperatures (20 – 23°C) in order to encourage the formation of lactic acid to the detriment of acetic acid.
desem	Natural leaven made from whole-wheat flour. Traditionally it is kept cool (18°C) and completely covered by flour in order to favor the development of the lactic acid bacteria and yeasts that are naturally present in the flour.

amish	Natural leaven or leaven made on the basis of commercial yeast in which, instead of water, milk is used. Sometimes a small amount of sugar is added as well.
Anfrischsauer	German expression, which indicates the first step in the production process of sourdough
Grundsauer	German expression used to indicate the second step in the production of sourdough. This process is described in detail later on.
Vollsauer	German expression used to indicate the third step in the production of sourdough. This process is described in detail later on.
madre (Spanish or Italian)	Natural leaven made with wheat flour that is allowed to ferment spontaneously.
poolish	Mixture of water, flour and commercial yeast. In most cases quite a bit of water is used so the final consistency resembles more a batter than dough.
Preferment	Is basically the same as a polish although certain preferment's are made with lactic acid bacteria (used in the production of crackers for instance).
Sauerteig	Leaven obtained by the natural and spontaneous fermentation of flour and water. After complete maturation, a piece is kept aside as "Anstellgut".
Sourdough	Leaven obtained by the natural and spontaneous fermentation of flour and water. After complete maturation a piece is kept aside to start the process all over again
starter (English or Dutch)	Can have two meanings pure culture of lactic acid bacteria freely available from specialized companies piece of dough which is kept aside and used as basis in a process of continuous refreshment.
sponge	Mixture of water, flour and commercial yeast. The consistency is similar to that of normal dough or slightly less solid.

Chemical Leaveners

Chemical leaveners are those that release gases produced by chemical reactions.

Baking Soda

Baking soda is the chemical sodium bicarbonate. If moisture and an acid are present, soda releases carbon dioxide gas, which leavens the product. Heat is not necessary for the reaction to take place (thoughthe gas is released faster at high temperatures). For this reason, products leavened with soda must be baked at once, or gases willescape and leavening power will be lost.

Acids that react with soda in a batter include honey, molasses, brown sugar, buttermilk, sour cream, yogurt, fruit juices and purées, chocolate, and natural cocoa (not Dutch-processed).

Sometimes cream of tartar is used for the acid. The amount of soda used in a formula is generally the amount needed to balance the acid. If more leavening power is needed, baking powder, not moresoda, is used.

Baking Powders

Baking powders are mixtures of baking soda plus one or more acids to react with it. They also contain starch, which prevents lumping and brings the leavening power down to a standard level.

Because baking powders do not depend for their leavening poweron acid ingredients in a formula, they are more versatile.

Baking Ammonia

Baking ammonia is a mixture of ammonium carbonate, ammonium bicarbonate, and ammonium carbamate. It decomposes rapidly during baking to form carbon dioxide gas, ammonia gas, and water. Only heat and moisture are necessary for it to work. No acids are needed.

Baking ammonia decomposes completely, leaving no residue that could affect flavour when it is properly used. However, it can be used only in small products that are baked until dry, such as cookies. Only in such products can the ammonia gas be completely driven off, because ammonia releases gases very quickly, it is sometimes used in products in which rapid leavening is desired, such as cream puffs. Use of ammonia enables the baker to lower the cost of such products by reducing the quantity of eggs. However, the quality of the resulting goods are lowered as well.

Storage of Chemical Leaveners

Baking soda, powder, and ammonia must always be kept tightly closed when not in use. If left open, they can absorb moisture from the air and lose part of their leavening power. They must be stored in a cool place, because heat also causes them to deteriorate.

Air

Air is incorporated into all doughs and batters during mixing. The formation of air cells is important even in products leavened by yeast or baking powder because the air cells collect and hold the leavening gases.

Some products are leavened mostly or entirely by air. In these products, air is incorporated into the batter primarily by two methods: creaming and foaming. This air expands during baking and leavens the products.

1. **Creaming** is the process of beating fat and sugar together to incorporate air. It is an important technique in cake and cookie making. Some pound cakes and cookies are leavened almost entirely by this method.
2. **Foaming** is the process of beating eggs, with or without sugar, to incorporate air. Foams made with whole eggs are used to leaven sponge

cakes, while angel food cakes, meringues, and soufflés are leavened with egg-white foams.

Steam

When water turns to steam, it expands to 1,100 times its original volume. Because all baked products contain some moisture; steam is an important leavening agent. Puff pastry, cream puffs, popovers, and piecrusts use steam as their primary or only leavening agent. If the starting baking temperature for these products is high, steam is produced rapidly and leavening is greatest.

SHORTENINGS

Shortenings

Any fat acts as a shortening in baking because it shortens gluten strands and tenderizes the product. However, we generally use the word shortening to mean any of a group of solid fats, usually white and tasteless, that are especially formulated for baking. Shortenings generally consist of nearly 100% fat.

Shortenings may be made from vegetable oils, animal fats, or both. During manufacturing, the fats are hydrogenated. This process turns liquid oils into solid fats. Shortenings are used for many purposes, so manufacturers have formulated different kinds of fats with various properties.

There are three main types: regular or all-purpose (AP) shortenings, high-ratio plastic shortenings, and high-ratio liquid shortenings.

Regular Shortenings

Regular shortenings, or all-purpose shortenings, have a fairly tough, waxy texture, and small particles of the fat tend to hold their shape in a dough or batter. They are called plastic shortenings, which means they are moldable at room temperature. Regular shortenings can be manufactured to varying degrees of hardness. They have a good creaming ability.

This means that a good quantity of air can be mixed into them to give a batter lightness and leavening power.

Also, this type of shortening melts only at a high temperature, because of their texture, regular shortenings are used for flaky products such as pie crusts and biscuits.

They are also used in many other pastries, breads, and products mixed by creaming, such as certain pound cakes, cookies, and quick breads.Unless another shortening is specified in a formula, regular shortening is generally used.

High Ratio Plastic Shortenings

These are soft shortenings that spread easily throughout a batter and quickly coat the particles of sugar and flour. They are called high-ratio because they were devised for use in cake batters that contain a high ratio of sugar and liquid to flour. They also contain added emulsifying agents, so they can hold a larger quantity of liquid and sugar than regular shortenings. Thus, they give a smoother and finer texture to cakes, and make them moister. Because of the added emulsifiers, this shortening is also commonly referred to as emulsified shortening.

On the other hand, high-ratio shortening does not cream well. When recipe instructions call for creaming shortening and sugar, regular shortening rather than high-ratio shortening should be used.

When emulsified shortening is used to make high-ratio cakes—cakes with a high ratio of sugar and liquid to flour—a simpler mixing method can be used because this shortening spreads so well (see Chapter 16). In addition, high-ratio shortening is often used in icings because it can hold more sugar and liquid without curdling.

The term emulsified shortenings is not, strictly speaking, an accurate one. Pure fat cannot be emulsified, because an emulsion is a mixture of at least two substances. It would, perhaps, be more accurate to call them emulsifier or emulsifying shortenings. However, the term emulsified shortenings is the more widely recognized and commonly used term.

High-Ratio Liquid Shortenings

High-ratio liquid shortenings, also called liquid cake shortenings, are less hydrogenated than plastic shortenings, making them liquid and pourable, although they are thick and cloudy or opaque in appearance. They contain more emulsifiers than high-ratio plastic shortenings, and are effective shortenings in high-ratio cakes. The emulsifiers make the cakes moist and fine textured. Also, because air is incorporated so easily during mixing, these shortenings increase the volume and tenderness of cakes.

Butter

Fresh butter in North America consists of about 80% fat, about 15% water, and about 5% milk solids. Most North American butter is made from sweet cream. Many European butters have a higher fat content—about 82%, or even more—and a lower moisture content. In addition, they are more likely to be made from cultured cream, which gives them a somewhat fuller flavor.

Butter is available salted and unsalted. Unsalted butter is more perishable, but it has a fresher, sweeter taste and is thus preferred in baking. Also, salt masks flavors that might be absorbed during storage, making it is harder to tell if salted butter has foreign flavours that might detract from the finished baked goods. If salted butter is used, the salt in the formula may have to be reduced. It is difficult to know for sure, however, how much to reduce the salt content, because salted butters vary in their composition.

Shortenings are manufactured to have certain textures and levels of hardness to suit them for particular uses. Butter, on the other hand, is a natural product that doesn't have this advantage. It is hard and brittle when cold, very soft at room temperature, and it melts easily. Consequently,

Dough made with butter are much harder to handle. Also, butter is more expensive than shortening.

On the other hand, butter has two major advantages:

1. **Flavor.** Shortenings are intentionally flavourless, but butter has a highly desirable flavour.
2. **Melting Qualities.** Butter melts in the mouth. Shortenings do not. After eating pastries or icings made with shortening, one can be left with an unpleasant film of shortening coating the mouth. For these reasons, many bakers and pastry chefs feel the advantages of butter outweigh its disadvantages for many purposes. Shortening is not often used in fine French pastries, for example. Frequently, you may blend 50% butter and 50% shortening to get both the flavour of butter and the handling qualities of shortening.

Margarine

Margarine is manufactured from various hydrogenated animal and vegetable fats, plus flavouring ingredients; emulsifiers; colouring agents; and other ingredients. It contains 80 to 85% fat, 10 to 15% moisture, and about 5% salt, milk solids, and other components. Thus, it may be considered a sort of imitation butter consisting of shortening, water, and flavouring.

Unlike the margarines sold by retail grocers, baker's margarines are formulated in different ways for different purposes. Following are the two major categories.

Cake and Baker's Margarines

These types of margarine are soft and have good creaming ability. They are used not only in cakes but also in a wide variety of other products.

Pastry Margarines

These margarines, also called roll-in compounds, are tougher and more elastic than cake margarines and have a waxy texture. They are especially formulated for doughs that form layers, such as Danish dough and puff pastry. Puff pastry margarine, the toughest of these fats, is sometimes called puff pastry shortening.

Puff pastry made with this margarine generally rises higher than pastry made with butter.

However, as the fat doesn't melt in the mouth like butter, many people find the pastry unpleasant to eat.

Roll-in margarine is somewhat softer in texture than puff pastry margarine and has a lower melting point. It can be used in Danish pastries, croissants, and puff pastry.

Oils

Oils are liquid fats. They are not often used as shortenings in baking because they spread through a batter or dough too thoroughly and shorten too much. Some breads and a few cakes and quick breads use oil as a shortening. Beyond this, the usefulness of oil in the bakeshop is limited primarily to greasing pans, deep-frying doughnuts, and serving as a wash for some kinds of rolls.

Lard

Lard is the rendered fat of hogs. Because of its plastic quality, it was once highly valued for making classic American flaky piecrusts and biscuits—and it is still sometimes used for these products. Since the development of modern shortenings, however, it is not often used in the bakeshop.

Storage of Fats

All fats become rancid when exposed to the air too long. Also, they tend to absorb odors and flavors from other foods. Highly perishable fats, such as butter, should be stored, well wrapped, in the refrigerator. Other fats and oils should be kept in tightly closed containers in a cool, dry, dark place.

GELLING AGENTS AND STABILIZERS

Gelling Agents and Stabilizers with their Role in Bakery and Confectionery

Gelling Agents

Gelling agents are food additives used to thicken and stabilize various foods, like jellies, desserts and candies.The agents provide the foods with texture through formation of a gel. Typical gelling agents include natural gums, starches, pectin's, agar-agar and gelatine.

Stabilizer

A stabilizer is an additive to food, which helps to preserve its structure. Uses include preventing oil, water emulsions from separating in products such as salad dressing; preventing ice crystals from forming in frozen food such as ice cream; and preventing fruit from settling in products such as jam, yogurt and jellies.

Most common stabilizers used alginate, agar, carrageen, gelatin, guar gum, pectin, starch, xanthan gum, cream of tartar

Some stabilizers and thickening agents are gelling agents.

Gelatine

Gelatine is a protein, used to thicken or to solidify (gel) soft or liquid foods. A water-soluble protein extracted from animal connective tissue. When a sufficient quantity of gelatin is dissolved in hot water or other liquid, the liquid will solidify when cooled or chilled, the strands bond with one another to form a network that traps the liquid to keep it from flowing at all.

When used in smaller quantities, the liquid will thicken but not solidify; the strands get tangled with one another, so the liquid does not flow as freely.

Gelatine thickens and gels because of the nature of its proteins, which form long strands.

Starch

Starch is used primarily to thicken puddings, pie fillings, and similar products. The three most important starches in dessert production are as follows:

1. **Cornstarch**

   It has a special property that makes it valuable for certain purposes. Products thickened with corn-starch set up almost like gelatine when cooled. For this reason, corn-starch is used to thicken cream pies and other products that must hold their shape.

2. **Waxy Maize**

   It is made from a different type of corn. It is usually manufactured into a form called modified food starch.

   Waxy maize and other modified starches do not break down when frozen, they are used for products that are to be frozen. They are clear when cooked and give a brilliant, clear appearance to fruit pie fillings.

   It does not set up firm like corn-starch but rather makes a soft paste that has the same consistency hot and cold. Thus, it is not suitable for cream pie fillings.

3. **INSTANT STARCHES**

   They are precooked or pregelatinized, so thicken cold liquids without further cooking. They are useful when heat will damage the flavour of product, as in fresh fruit glazes such as strawberry.

Amylose and Amylopectin Starches

Starch molecules fall into two principal categories.

Amylose Starches

Amylose molecules are long, straight chains molecules,most grain starches are high in these starches.Amylose starches, after cooking, get thicker and cloudier as they cool,they form a firm gel when cooled. They tend to break down and release liquid after long storage or after freezing.

Amylopectin Starches

Amylopectin molecules have many branches. Starches from roots and tubers, such as potatoes and arrowroot, are high in amylopectin starches.Amylopectin starches do not get thicker as they cool, and they remain fairly clear.

They also remain stable when frozen and do not release liquid in storage or after freezing.

Pectin

Pectin is a soluble fibre, a type of carbohydrate that is not absorbed by the body. It is present in many fruits, but most commercial form of pectin is extracted from peels of citrus fruits.In general, unripe fruits have more pectin than ripe fruits. One of the reasons fruits get softer as they ripen is that the pectin breaks down.

Pectin is used to thicken or gel fruit preserves, jams, and jellies. It can also be used to make fruit glazes, because the pectin thickens or sets fruit juices and purées.

An important advantage of pectin over an ingredient such as corn-starch is that it makes a clear, not cloudy, gel.Pectin alone, when added to liquid, thickens but does not solidify or gel. For pectin to gel, an acid (such as fruit juice) and a high sugar content are necessary, so reason jams & jellies contain so much sugar.

Fruits high in pectin will gel naturally, without needing additional pectin. Such fruits include cranberries, apples, and plums. Citrus peels also contain pectin; this characteristic is useful in the production of marmalade.

For making jellies, the amount of pectin you need to add varies with the fruit. As a rule of thumb, 1.75 ounces (50 g) powdered pectin gels 2 quarts (2 L) fruit juice or fruit pulp.

Vegetable Gums

Vegetable gums are carbohydrates that consist of molecules in long chains, somewhat like starches.They can absorb a great deal of water, which makes them useful for thickening or gelling liquids.

Most of these gums are used by manufacturers and are not likely to be found in the bakeshop. The following three gums, however, are sometimes used by pastry chefs and retail bakers

Agar-Agar

Also known simply as agar or kanten (its Japanese name) is derived from seaweed and sold in the form of dry strands or powder. It is used much like gelatine, except it does not need to be refrigerated in order to gel. This makes it especially good for use in warm weather and for products that will not be refrigerated.

Unlike gelatine, which is an animal product, agar-agar can be used on vegetarian menus. As a rule of thumb, 1 part agar by weight has as much gelling power as 8 parts gelatine.

Gum Tragacanth

Often called gum trag, is extracted from a bush native to the Middle East. It is used by pastry chefs to make gum paste, a decorative product similar to pastillage.

XANTHAN GUM (ZAN than)

Often used to provide structure in gluten-free formulas.

Food Additives and Preservatives

Food Additives

Food additives are substances added to food to maintain or improve the safety, freshness, texture, to preserve flavor or enhance its taste, appearance, or other qualities. Some food additives have been in use for centuries for preservation – such as salt (in meats such as bacon or dried fish), sugar (in marmalade), or sulfur dioxide (in wine).

Food additives need to be checked for potential harmful effects on human health before they can be used. Food additives can be derived from plants, animals, or minerals, or they can be synthetic. They are added intentionally to food to perform certain technological purposes.

There are several thousand food additives used, all of which are designed to do a specific job in making food safer or more appealing.

Some food additives available for bakery are Calcium stearoyl lactylate, Sodium stearoyl lactylate, Sodium stearoyl fumarate, Potassium bromate, Carrageenan, Tetrasodium pyrophosphate, Fumaric acid.

Food Preservatives

Food preservatives are employed to ensure safety and avoid quality loss derived from microbial, physical–chemical, or enzymatic reactions. They are used in baking to extend shelf life and preserve quality characteristics during transport and commercialization. Food preservatives are both natural and artificial.

The three main categories of artificial preservatives are

1. **Antimicrobials**: restrict and inhibit the growth of spoilage organisms, such as mold and rope bacteria.
2. **Antioxidants**: stabilize fats and oils, as well as delay their oxidation. In addition, they prevent discoloration of natural pigments.
3. **Chelating agents (S**equestrants): aid in binding metals that can cause oxidation in fats and oils.

Some commonly used **Artificial preservatives** are **Benzoates, Butylated hydroxyanisole, Polyphosphates**

Some commonly use **Natural preservatives** are **Vinegar, Vitamin C, Cultured Starches, Spices, Plant extracts**

Edible Garnishes

A garnish is an item or substance used as a decoration or embellishment accompanying a prepared food dish or drink. In many cases, it may give added or contrasting flavor. This decorative item can add important flavor, texture, and functional elements to the plating of the dessert, and can enhance the enjoyment of the dish.

Some garnishes are selected mainly to augment the visual impact of the plate, while others are selected specifically for the flavour they may impart. A food item which is served with garnish may be described as being garni, the French term for garnished.

Garnishes can solve the problem of serving a frozen component (ice cream or sorbet, for example) as part of a plated dessert. If a scoop of ice cream is placed directly onto the plate, it will start melting immediately, marring the presentation. If that scoop is placed onto an item, such as a cookie base, it will slow down the rate of melting, making it easier to serve, and allowing the customer to enjoy and appreciate the effort spent on the presentation.

Other garnishes that can be used to help present a frozen component are:

1. Tuiles, which will also add visual, flavor, and texture (crunchy) elements
2. Sliced fruit
3. Meringue disk
4. Small cookie
5. Chocolate garnish
6. Crumbs (cake, crushed nuts, or brittle)

Some of the more popular items that can be used as garnishes are described below.

Tuiles

These thin cookies can be shaped in numerous ways: with a stencil, spread onto a silicone baking sheet; combed; piped; or spread onto a textured flexible baking mat and then shaped while still warm.

Meringues (Hard)

When making meringues, keep the ratio of sugar to egg whites 1.5-2 to 1. Meringues can be made with the common or Swiss technique, and must be dried in a low-temp oven for 12 to 36 hours. Meringue can be piped into sticks, disks, or baskets such as those made to produce the classic French dessert vacherin.

Sugar

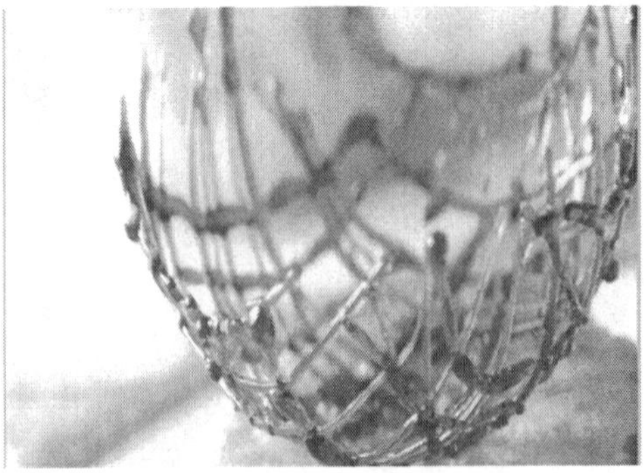

Sugar can be used to make garnishes using several different methods. Pulled sugar is prepared by cooking a sugar syrup and then cooling, coloring, folding, and stretching it into various shapes such as ribbons or bows.

The same mixture can also be used to make blown sugar garnishes, which are created by using a pump to create a balloon out of the slightly cooled sugar syrup. Blown sugar is used to make things such as fruits, with many other possibilities. Bubble sugar is made by pouring liquid sugar onto a special type of paper and then lifting the paper and allowing it to run down, creating a bubbly effect.

Caramelized Sugar

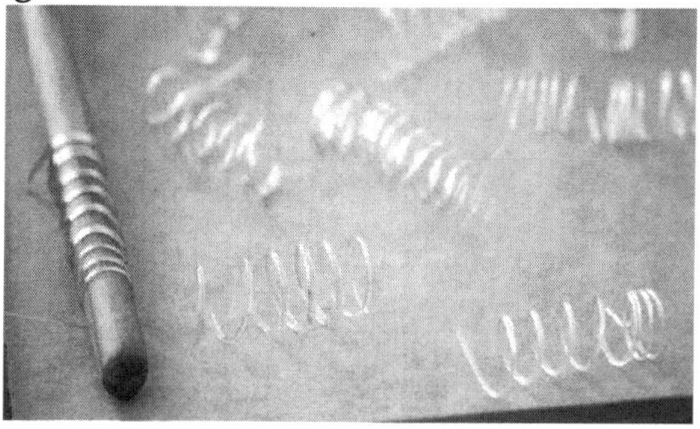

It can be used to make spun sugar, can be piped into shapes or sticks or drizzled over the back of a bowl to make a caramel cup, or nuts can be dipped into the caramel and pulled to create spikes.

Isomalt Sugar

It can be used much like regular sugar to create pulled and blown sugar, or sprinkled between two silicone baking sheets and baked at 175°C (350°F) for 12 to 15 minutes. For additional effect, it can be dusted with powdered food colours before baking.

Chocolate

Chocolate has many different applications. Too thick a chocolate garnish can overpower the dessert, so it must be delicate. In all cases, chocolate must be correctly tempered, which will ensure a crisp texture and proper sheen.

Some garnishes that can be prepared are chocolate curls, fans, and cigarettes, formed by spreading a thin layer of tempered chocolate onto a marble slab and then shaping once partially set.

Chocolate can also be piped into shapes, mixed with nuts and poured to form bark, or mixed with cream and used for specification to create chocolate caviar.

Marzipan/Rolled Fondant

Marzipan and fondant can be rolled and cut out into shapes and figurines, or used for bases.

Dough and Pastry

Filo pastry

Filo can be buttered and layered, with flavors, such as nuts, seeds, cocoa powder, herbs, and spices, added between layers. It can also be cut into shapes and made into cups, etc., and baked.

Kataifi dough

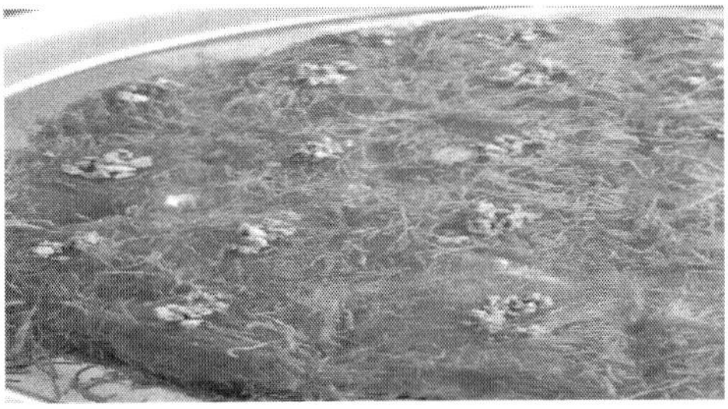

Similar to a filo pastry but in thin strands, kataifi dough is commonly brushed with butter before being baked. Can be tied into knots or baskets and baked off. Found in Middle Eastern cuisines.

Bric dough

Bric comes in sheets and is brushed with water, baked with cinnamon sugar, and cut to make interesting shapes. Found in Middle Eastern cuisines.

Puff pastry

Puff pastry can be rolled into sheets and used as a base, cut and twisted to form straws and allumettes, etc. Puff pastry also adds a textural component to desserts.

Choux paste

Choux paste can be piped or combed into long strands and dusted with seeds or nuts before baking.

Cheese

Sprinkle grated hard cheese carefully onto a baking sheet and bake until crisp, approximately 5 to 10 minutes at 175°C (350°F). The pieces can be broken into shards when cool.

Berries and Fruit

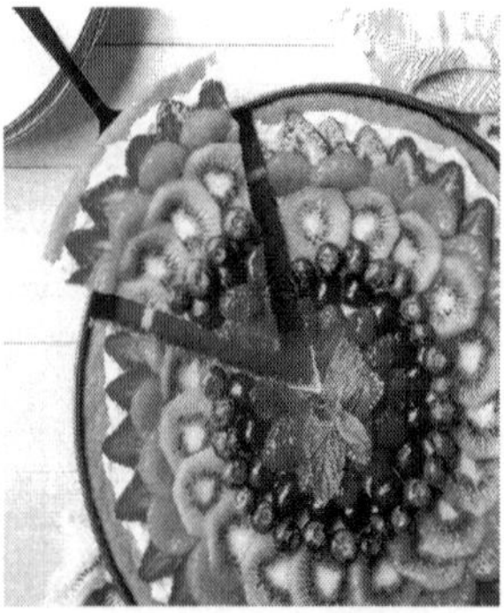

Berries and other fruits can be used fresh or dried to make fruit leather or powders.

Dried fruit

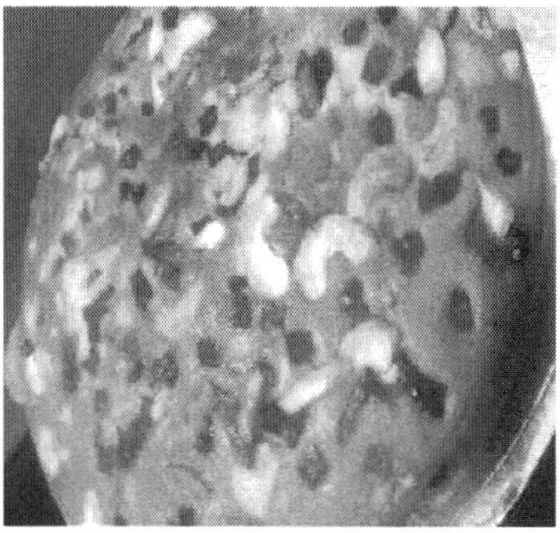

Firm fruits can be sliced thinly and soaked in sugar syrup with lemon juice briefly before drying in a low-temperature (95°C or 200°F) oven on a silicone baking sheet for several hours. Fruit treated the same way can also be dried in a dehydrator.

Candied zest

Citrus peels can be poached in sugar syrup and then cooled and coated in finely granulated sugar.

Candied Nuts and Brittles

Heat sugar syrup to the hard crack stage (148°C to 155°C or 310°F to 330°F) and add toasted nuts.Remove, drain excess syrup, and cool. Nuts can also be chopped and cooked in a syrup to form a nut brittle, which can be broken into pieces or ground to make a powder.

Rice Noodles

Rice or other noodles can be deep-fried briefly and tossed in cinnamon sugar.

Cookies

A wide range of cookies, such as sugar cookies, shortbreads, etc., can be used as garnishes. Cookies can also be made from sweet dough (pate sucré), tart dough, etc.

Sponges

Different types of cakes and sponges, such as genoise, angel food, japonaise, joconde, and baumkuchen, can be sliced thinly and cut into different shapes. Caviar and other interesting garnishes can be made with the reverse spherification method.

Canned Food

Canned foods are a significant component of the diet of most individuals in developed countries, offering food in a convenient form with year-round availability. The canning process relies on heat treatment for the destruction of microorganisms and preservation of the food, which is then generally considered to have an indefinite microbiological shelf life providing that pack integrity be maintained.

Canning is a method of preserving food in which the food contents are processed and sealed in an airtight container (jars like Mason jars, and steel and tin cans).The extent of the thermal processing, in terms of both temperature and duration of the treatment, is dependent upon the chemical and physical composition of the product. Both physical and chemical changes occur during processing and, to a lesser extent, during storage, and it is these that determine the product quality in terms of its sensory properties and nutrient content.These changes, which can be either desirable or undesirable, are influenced by the time and temperature of the process, the composition and properties of the food, the canning medium, and the conditions of storage.

Canning provides a shelf life typically ranging from one to five years, although under specific circumstances it can be much longer.

A freeze-dried canned product, such as canned dried lentils, could last as long as 30 years in an edible state.

Commonly used canned foods include all types of fruits and vegetables, milk, coconut milk, chickpeas, corn, prunes, sauces, olives, Jalapenos

Introduction to Food Laws

- The Indian food processing industry is regulated by several laws, which govern the aspects of sanitation, licensing and other necessary permits that are required to start up and run a food business.
- The legislation that dealt with food safety in India was the Prevention of Food Adulteration Act, 1954 (hereinafter referred to as "PFA"). The PFA

had been in place for over five decades and there was a need for change due to varied reasons, which include the changing requirements of our food industry.

- The act brought into force in place of the PFA is the Food Safety and Standards Act, 2006 (hereinafter referred to as "FSSA") that overrides all other food related laws. It specifically repealed eight laws which were in operation prior to the enforcement of FSSA:
- The Prevention of Food Adulteration Act, 1954
- The Fruit Products Order, 1955
- The Meat Food Products Order, 1973
- The Vegetable Oil Products (Control) Order, 1947
- The Edible Oils Packaging (Regulation) Order, 1998
- The Solvent Extracted Oil, De oiled Meal, and Edible Flour (Control) Order, 1967
- The Milk and Milk Products Order, 1992
- Essential Commodities Act, 1955 (in relation to food)

Storage and Preservation of Bakery and Confectionery Products

Storage of Fats

- All fats become rancid when exposed to the air too long.
- They tend to absorb odours and flavours from other foods.
- Highly perishable fats, such as butter, should be stored, well wrapped in the refrigerator.
- Other fats and oils should be kept in tightly closed containers in a cool, dry, dark place.

Storage of Milk Products

- Fresh milk and cream, buttermilk and other fermented milk products, and cheese must be kept refrigerated at all times.
- Evaporated milk in unopened cans may be kept in a cool storage area. After opening, however, it must be stored in the refrigerator.
- Condensed milk in large containers keeps for a week or more after opening if kept covered and in a cool place. The sugar acts as a preservative. Stir before using because the sugar tends to settle to the bottom and sides.
- Dried milk does not need refrigeration, but should be kept in a cool, dark place, well away from ovens and other heat sources. Keep the container tightly closed, to prevent the milk from absorbing moisture from the air.

Storage of Egg

- Proper storage is essential for maintaining egg quality.
- Eggs keep for weeks when held at 36°F (2°C) but lose quality quickly when held at room temperature.

- In fact, they can lose a full grade in one day at warm bakeshop temperatures.
- There is no point in paying for Grade AA eggs if they are Grade B by the time you use them.
- Store eggs away from other foods that might pass on undesirable flavors or odors.
- Storage of eggs in chiller until they are needed. When eggs are required to store at room temperature, current batch of eggs to be used with daily replenishment of stock.

Storage of Chemical Leaveners

- Baking soda, powder, and ammonia must always be kept tightly closed when not in use.
- If left open, they can absorb moisture from the air and lose part of their leavening power.
- They must be stored in a cool place, because heat also causes them to deteriorate.
- Storage of dry ingredients like wheat flour, sugar, milk powders, minor ingredients and cocoa powder
- All dry ingredients should be stored in airtight containers in cool and dry place.
- Flour storage in freezer is the first best place, then refrigerator after that room with controlled temperature. Flour should be stored in airtight containers in a cool, dry place (less than 60% humidity).
- All purposes, bread and cake flour, will keep for 6 months to a year at 70% and 2 years at 40%.
- Store away from foods with strong odours.
- They should be stored in a room that has the required humidity and temperature control.
- The enemy of flour storage is light — both because it generates heat, and because light itself can encourage oxidation, which is the enemy of fresh flour. Therefore, flour storage shall be dark.
- Whole-wheat or whole-grain flours have more of the natural oils that can spoil, so discard after one month at regular room temperature, three months at cooler house temperature, six months in fridge, or one year in the freezer.

Nuts are stored below 4°C to assure dormancy of the insect eggs, if any in the nuts.

Storage of finished products in covered containers in the chiller at 4°C and below.

Storage of finished products on separate shelves above raw food (including shell eggs).

Organic and Genetically Modified Foods

Oraganic Food

The term organic refers to the way agricultural products are grown and processed. While the regulations vary from country to country, in the U.S., organic crops must be grown without the use of synthetic pesticides, bioengineered genes (GMOs), petroleum-based fertilizers, and sewage sludge-based fertilizers.

Organic livestock raised for meat, eggs, and dairy products must have access to the outdoors and be given organic feed. They may not be given antibiotics, growth hormones, or any animal by-products.

Organic Vs Non Organic

Organic Produce	Conventionally-Grown Produce
• Grown with natural fertilizers (manure, compost)	• Grown with synthetic or chemical fertilizers
• Weeds are controlled naturally (crop rotation, hand weeding, mulching, and tilling)	• Weeds are controlled with chemical herbicides.
• Pests are controlled using natural methods (birds, insects, traps) and naturally-derived pesticides.	• Pests are controlled with synthetic pesticides
Organic Meat, Dairy, Eggs:	**Conventionally-Raised Meat, Dairy, Eggs**
• Livestock are given all organic, hormone- and GMO-free feed.	• Livestock are given all organic, hormone- and GMO-free feed.
• Disease is prevented with natural methods such as clean housing, rotational grazing, and healthy diet.	• Antibiotics and medications are used to prevent livestock disease.
• Livestock must have access to the outdoors.	• Livestock may or may not have access to the outdoors.

Benefits of Organic Food

- The way in which food is grown or raised can have a major impact on your mental and emotional health as well as the environment.
- Organic foods often have more beneficial nutrients, such as antioxidants, than their conventionally grown counterparts and people with allergies to foods, chemicals, or preservatives often find their symptoms lessen or go away when they eat only organic foods.

Organic produce contains fewer pesticides. Chemicals such as fungicides, herbicides, and insecticides are widely used in conventional agriculture and residues remain on (and in) the food we eat. They only use naturally derived pesticides, rather than the synthetic pesticides used on conventional commercial farms. Natural pesticides are believed to be less toxic, however, some have been found to have health risks.

Organic food is often fresher because it doesn't contain preservatives that make it last longer. Organic produce is often (but not always, so watch where it is from) produced on smaller farms near where it is sold.

Organic farming is better for the environment. Organic farming practices reduce pollution, conserve water, reduce soil erosion, increase soil fertility, and use less energy. Farming without pesticides is also better for nearby birds and animals as well as people who live close to farms.

Organically raised animals are not given antibiotics, growth hormones, or fed animal byproducts. Feeding livestock animal byproducts increases the risk of mad cow disease (BSE) and the use of antibiotics can create antibiotic-resistant strains of bacteria. Organically-raised animals are given more space to move around and access to the outdoors, which help to keep them healthy.

Organic meat and milk are richer in certain nutrients. Results of a 2016 European study show that levels of certain nutrients, including omega-3 fatty acids, were up to 50 percent higher in organic meat and milk than in conventionally raised versions.

Organic food is GMO-free. Genetically Modified Organisms (GMOs) or genetically engineered (GE) foods are plants whose DNA has been altered in ways that cannot occur in nature or in traditional crossbreeding, most commonly in order to be resistant to pesticides or produce an insecticide.

Genetically Modified Foods

Genetically modified (GM) foods are foods derived from organisms whose genetic material (DNA) has been modified in a way that does not occur naturally, e.g. through the introduction of a gene from a different organism.

The technology is often called "modern biotechnology" or "gene technology", sometimes also "recombinant DNA technology" or "genetic engineering".

Currently available GM foods stem mostly from plants, but in the future foods derived from GM microorganisms or GM animals are likely to be introduced on the market.

In the future, genetic modification could be aimed at altering the nutrient content of food, reducing its allergenic potential or improving the efficiency of food production systems.

All GM foods should be assessed before being allowed on the market. FAO/WHO Codex guidelines exist for risk analysis of GM food.

Effects of GM Foods on Human Health

All genetically modified foods currently available on the international market have passed safety assessments and no effects on human health have been shown as a result of consuming GM foods.

When discussing and developing GM foods, three main safety issues are considered.

Allergenicity, meaning the ability of the edited genes or food product to cause an allergic reaction. No allergic effects have been found relative to GM foods currently on the market.

Transfer of Genes from GM foods to the human digestive tract. The probability of transfer is low; however, the use of gene transfer technology that does not involve antibiotic resistance genes is encouraged.

Outcrossing, meaning the transfer of genes from GM organisms to other species. Several countries have developed preventative strategies, including clear separation of GM and non-GM food crops.

Reasons to Produce GM Foods

- GM foods can also allow for reductions in food prices through improved yields and reliability.
- Most existing genetically modified crops have been developed to improve yield through the introduction of resistance to plant diseases or of increased tolerance of herbicides.
- Resistance against insects is achieved by incorporating into the food plant the gene for toxin production from the bacterium Bacillus thuringiensis (Bt).
- The toxin is currently used as a conventional insecticide in agriculture and is safe for human consumption.
- GM crops that inherently produce this toxin have been shown to require lower quantities of insecticides in specific situations, e.g. where pest pressure is high.
- Virus resistance is achieved through the introduction of a gene from certain viruses which cause disease in plants.
- Virus resistance makes plants less susceptible to diseases caused by such viruses, resulting in higher crop yields.
- Herbicide tolerance is achieved through the introduction of a gene from a bacterium conveying resistance to some herbicides.
- In situations where weed pressure is high, the use of such crops has resulted in a reduction in the quantity of the herbicides used.
- Commonly grown GM foods are Corn, Soy, Yellow Crookneck Squash and Zucchini, Alfalfa, Canola, Sugar Beets, Milk.

3. OVEN AND BAKING

Oven Temperature

Oven temperature is one of the key baking parameters.

It can be measured, modified, and controlled in order to influence process conditions directly, thereby affecting a product's final characteristics.

Importance of Oven Temperature

Oven temperature causes physical transitions and chemical reactions to take place in the dough/batter.The following stages are temperature-dependent, and participate in the sequential transformation of bread dough.

1. Development (Also Known As Oven Spring)

As temperature increases, the free water/alcohol mixture in the product vaporizes, fermentation gases (CO2) dissolved in the liquid dough phase become less soluble, and are released into the cells, causing them to expand in response to the rise in pressure.

As a result, cells increase in volume by retaining gases due to their deformable nature as these are surrounded by the continuous gluten/starch soft matrix. This results in a large reduction in the density of the dough as the product gradually develops an aerated structure.

Note that the size to which the gas bubbles can grow is limited by the ability of the gluten/starch film surrounding them to stretch without rupturing.

In this stage, the product undergoes a series of irreversible chemical and physical transformations. Oven spring is accompanied by the following changes and conditions:

Killing of yeast cells at 50–60°C (122–140°F)

Maximum enzymatic activity at 60°C (140°F). The enzyme-driven reactions that convert starch into sugars and break proteins into amino acids increase with heat, so they increase most near the dough surface.

Starch gelatinization: It starts at 55–65°C (130–150°F) as granules become fully swollen with local free water.

Denaturation of gluten proteins at 50°C (122°F) and coagulation at 70–80°C (160–180°F). As a consequence, gluten becomes increasingly tough and stiff as it irreversibly forms a gel.

Above 85°C (185°F), starch looks glassy, and gluten looks rubbery. This is the start of the dough-crumb transition process (setting).

Inactivation of naturally-occurring and added enzymes inside the dough (70–85°C) (160–185°F).

2. Drying (Reduction Of Dough/Batter Moisture)

Under the action of the heat transfer mechanisms, high temperatures develop inside the baking chamber (200–300°C) (390–570°F), and water molecules at the dough surface absorb latent heat and start to evaporate.

Due to the low humidity of the air inside the baking chamber, a water vapor pressure (air moisture concentration) gradient is created. Liquid-state water starts to diffuse, and migrates from the product core to the surface, where it evaporates and is lost to the oven atmosphere.

The loss of moisture from the dough piece is dependent on the baking chamber temperature, colligative properties of the free water in the product, heat transfer methods used, and the humidity of the oven.

3. Color Formation

The external surface of the product is exposed directly to the high temperatures of the oven, and readily absorbs the heat from the energy sources.

These high temperatures trigger non-enzymatic reactions that give rise to the desirable brown crust.

Maillard browning takes place above approximately 105°C (220°F) and requires the presence of a reducing sugar (glucose, maltose, or lactose) together with an amino acid, the type of which determines color and flavor.

Sugars caramelize at 160°C (320°F). This reaction will happen only in the presence of water.

Baking Temperatures and Time

	Temp (F/C)	Minutes
BREADS		
Biscuits	425 - 450 F 218 - 232 C	10 - 15
Cream Puffs	375 F 190 C	60
Popovers	375 F 190 C	60
Quick Loaf Breads	350 - 375 F 177 - 190 C	60 - 75
Yeast Bread	400 F 205 C	30 - 40
YEAST ROLLS		
Plain	400 - 425 F 205 - 218 C	15 - 25
Sweet	375 F 190 C	20 - 30
CAKES WITH FAT		
Cupcake	350 - 375 F 177 - 190 C	15 - 25
Layer Cake	350 - 375 F 177 - 190 C	20 - 35
Loaf Cake	350 F 177 C	45 - 60
Cakes Without Fat		
Angel Food & Sponge	350 F 177 C	50 - 60
Cookies		
Drop	350 - 400 F 177 - 205 C	8 - 15
Rolled	375 F 190 C	8 - 10

Egg, Meat, Milk & Cheese		
Soufflé (in a hot water pan)	350 F 177 C	30 - 60
Custard (in a hot water pan)	350 F 177 C	30 - 60
Meat Loaf	350 F 177 C	60 - 90
Meat Pie	400 F 205 C	25 - 30
Rice Pudding (raw rice)	300 F 149 C	120 - 180
Scalloped Potatoes	350 F 177 C	60
Pastry		
1 Crust Pie (Custard Type)	400 - 425 F 205 - 218 C	30 - 40
Shell Only	450 F 232 C	10 - 12
2 Crust Pies with Uncooked Filling	400 - 425 F 205 - 218 C	45 - 55
2 Crust Pies with Cooked Filling	425 - 450 F 218 - 232 C	30 - 45

4. BREAD

Functions of Ingredients in Bread Making and their Role

Introduction

A bread formulation contains many ingredients to suit the specific requirements of the variety of bread and the bread making method used. These ingredients, which are included in the bread formulation, are generally classified into essential and optional ingredients. The essential ingredients are flour, water, salt and yeast. The optional ingredients that are often found in the formula are fat, sweetener, malt, surfactants, milk and milk products, oxidizing agents, enzyme preparations, vital wheat gluten, yeast foods, anti-mold agents etc. Each of these compounds in the bread formula performs a specific function in producing a loaf of bread with desired quality attributes.

Wheat Flour

Wheat flour is the major component in a bread formulation and irresponsible for the structure of bread. Flour when mixed with water allows the formation of a viscoelastic dough that retains gas during fermentation and proofing stages. This property of flour is mainly due to the presence of endosperm proteins, namely, the gliadin and the glutenin. These proteins are unique, in that, when flour is wetted and mixed into a dough, the gliadin and glutenin proteins interact with each other to from the gluten proteins. In addition, the gluten proteins impart the visco-elastic properties, to wheat flour dough. The viscoelastic property of the wheat flour dough helps it to retain the carbon dioxide gas produced during the bread-making process and develop a honeycomb structure inside the baked loaf.

Based on their suitability for the manufacture of yeast-leavened bread, common wheats and flours milled from them are classified broadly into two groups: strong and weak.

Strong and weak denotes the strength of wheat relating to its baking quality. Some of the differences seen between strong and weak flours are:

1. Strong flour contains relatively high percentage of gluten forming proteins. On the other hand, weak flour has relatively low protein content.
2. Proteins form strong flour form tenacious, elastic gluten of good gas retaining property. Proteins from weak flour forms a weak, relatively, more extensible gluten of poor gas retaining capacity.
3. Strong flours require more water to make a dough of proper consistency. Weak flours has relatively low water absorption.
4. Because of high water absorption strong flour can give a high yield of bread. A weak flour gives lower bread yield.
5. The dough made from strong flour has excellent handling qualities and off more flexibility in mixing and fermentation requirements. For this reason, with hard wheat flour, we can produce bread over a wide range of baking conditions. Contrarily, dough from weak flours have poor fermentation tolerance, so they are more likely to fail in baking. It also has critical mixing and fermentation requirements. Weak flours require less mixing and fermentation than strong flours to yield optimum results.
6. Bread baked from strong flours have a well-risen loaf, good crumb grain and softer texture. Weaker flours give bread with poor loaf volume, crumb grain and poor texture.
7. Strong flours has the ability to carry a proportion of weak flour or flours of grains such as soya without affecting the quality of the final product.

Hence, wheat's bread making potential is derived largely from the quantity and quality of the gluten proteins. However, it has also been well recognized that practically all components of wheat flour are essential in one way or the other to produce an acceptable loaf of bread.

Water

Water is needed in the formation of dough having viscoelastic properties. The consistency of the dough is related to its water content. Water is needed to form the gluten proteins and for the necessary swelling of starch. It has been estimated that about 45.5% of the total water absorbed is associated with the starch, 31.2% with proteins, and 23.4% with the pentosans.

Water also acts, as solvent in the dough and many of the reactions that take place during fermentation cannot occur if there is no solvent. For example; water acts as a solvent for some of the released carbon dioxide gas to form carbonic acid; Carbonic acid contributes to the acid pH of the dough during fermentation providing a feasible atmosphere for the action of enzymes and yeast in the dough system. Water also facilitates the dissolution and uniform distribution of sugar, malt, milk etc. throughout the dough mass.

Quality of water namely, hardness and pH of water, plays an important role in bread making.

Hardness in water is due to the presence of calcium and magnesium ions. Calcium and magnesium can be present in the form of their bicarbonates, chlorides, sulphates, phosphates etc. Hardness is expressed in ppm of Calcium carbonate equivalent to all the calcium, magnesium and other constituents that contribute to hardness.

Hardness of water is classified into five groups. 0-5 very soft water; 15-50 soft water; 50-100 medium hard water; 100-200 hard water; 200 and above, very hard water. A water of hardness not exceeding (100 ppm) is most suitable for use in baking.

Some of the mineral salts found in hard water has strengthening effect on gluten. To some extent, these salts also serve as yeast food. Excessively hard water is undesirable because it retards fermentation by tightening the gluten structure too much. Soft water, on the other hand, lacks the gluten strengthening minerals and hence yield soft and sticky doughs.

Apart from the hardness of water, pH of the water also plays an important role in bread making. Excessive amounts of calcium and magnesium in the form of their bicarbonates are also responsible for the high alkalinity. Alkaline waters are undesirable because they tend to increase the dough pH. Doughs made with alkaline waters tend to show abnormal fermentation behavior because the pH level remains above the optimum range for yeast and enzyme activity.

Salt (Sodium Chloride)

Salt performs three principal functions in bread dough. Most important of these is the improvement of flavor in bread. Bread from which salt has been omitted is. insipid and flat in taste and flavor and is normally unsalable.

Second function of salt is its inhibiting effect on fermentation which is reflected in reduced yeast activity and hence, in a lower gassing power. This action of salt may prove desirable in instances of inadequate temperature control, where fermentation would otherwise proceed at an erratic rate leading to non-uniform results.

A third function of salt is its strengthening and tightening effect on the gluten of dough, which may be due in part to its inhibitory effect on proteolysis enzymes. This strengthening effect may prove especially desirable in instances where very soft water is encountered of where inadequately matured flours must be processed. Under these conditions, use of maximum amounts of salt will help overcome possible difficulties with soft and sticky dough.

While salt is an essential bread ingredient, not all its effects during the production process are considered desirable. For example, salt has a definite inhibiting effect on yeast activity, which is reflected in a reduction of the amount of carbon dioxide generated by the yeast within a given time. This is advantageous under circumstances where high dough temperatures prevail, but undesirable where a rapid fermentation under proper temperature control is aimed at. The fact that the presence of salt in dough markedly extends the time required to attain optimum mixing has led to the general practice of delayed salt addition. The extension of mixing time may be attributed partly to a salt-included reduction in water-holding capacity of the flour proteins, thereby increasing the dough "free water" content, and partly to the inhibiting effect of salt on proteolysis enzyme action. These two effects combine to produce slacker dough at normal absorption levels that requires a more extended mixing time.

Attempts to overcome these undesirable salt effects have resulted in the development of a special coated salt in which the individual salt particles are enclosed in a continuous film of special fat. The fat employed for this purpose has a relatively high melting point so that the coating will remain intact at the temperatures encountered during mixing and fermentation, but will release the salt at the higher proofing and baking temperatures. Such coated salt yields dough with characteristics that resemble those of salt-free dough, yet provides the desired flavor enhancement in the finished baked product.

Baker's Yeast

Baker's yeast is a unicellular microorganism belonging to the fungi group. It generally belongs to the species. Saccharomyces cerevisiae but some other species are also known to be capable to leavening bread.

Baker's yeast can be grouped into two main types:

1. Compressed yeast, also called "fresh yeast" and 2.Dried yeast.

Compressed Yeast

It is the traditional formulation of baker's yeast. It is sold as blocks of yeast cake wrapped in wax paper or cellophane.

Compressed yeast has a dry matter content of about 27-34% (w/w) and a protein content of about 42-56% (Dry weight basis). It is ready to be used immediately and, if it has been handled properly, it should give good result in all types of dough systems.

This type of yeast is perishable and should be stored at low temperatures, preferably between 0 and 4°C. At these temperatures, a shelf life of 3-4 weeks is possible with only slight decrease in leavening capacity. At higher temperatures, keeping quality is progressively lower. For frozen dough applications, it is often recommended to use compressed yeast as fresh as possible.

Dried Yeast

It is available in two main forms: Active Dry Yeast (ADY) and instant dry yeast (JOY). ADY, dates back to the 1920s but is was not until after World War II that its commercial production developed substantially. ADY is normally sold in airtight packages, vacuum-sealed or filled with an inert gas such as nitrogen. The leavening activity of ADY is substantially less than that of freshly used compressed yeast. However, the shelf life of ADY in a sealed package at room temperature is at least one year.

Bread Making Procedure

Although all yeast doughs are made according to the same basic principles, it is useful to divide yeast products into the three categories defined in this section.

Lean Dough Breads

A lean dough is one that is low in fat and sugar. Products made from lean doughs include the following:

Hard-crusted breads and rolls, including French and Italian breads, kaiser rolls and other hard rolls, and pizza. These are the leanest of all bread products.

Other white and whole wheat breads and dinner rolls. These have a higher fat and sugar content and sometimes also contain eggs and milk solids. Because they are slightly richer, they generally have soft crusts.

Breads made with other grains. Rye breads are the most common. Many varieties of rye bread are produced, with light or dark flours or with pumpernickel flour, and with various flavourings, especially molasses and caraway seeds.

Rich Dough Products

There is no exact dividing line between rich and lean doughs but, in general, rich doughs are those that contain higher proportions of fat, sugar, and sometimes eggs. Among the products made from rich doughs are:

Nonsweet breads and rolls, including rich dinner rolls and brioche. These have a high fat content, but their sugar content is low enough to allow them to be served as dinner breads. Brioche dough, made with a high proportion of butter and eggs, is especially rich.

Sweet rolls, including coffee cakes and many breakfast and tea rolls. These have high fat and sugar content and usually contain eggs. They generally have a sweet filling or topping.

Laminated or Rolled-In Yeast Dough Products

Rolled-in doughs or laminated doughs are those in which a fat is incorporated into the dough in many layers using a rolling and folding procedure. The alternating layers of fat and dough give the baked product a flaky texture.

Laminated doughs vary in sugar content from about 4% for some croissant doughs to 15% or more for some Danish doughs. However, most of the sweetness of laminated yeast dough products comes from the fillings and toppings.

Croissant and Danish doughs are the main laminated yeast dough products. In general, Danish dough products contain eggs, while croissants do not, although there are exceptions to this rule.

Bread Making Steps and Mixing Methods

There are12 basic steps in the production of yeast breads. These steps are generally applied to all yeast products, with variations depending on the particular product. In particular, many of the handcrafted artisan breads that have become popular require more complex procedures. The 12 steps are listed here.

1. Scaling ingredients
2. Mixing
3. Bulk fermentation
4. Folding
5. Dividing (scaling or portioning of dough)
6. Rounding
7. Bench-proofing or intermediate proofing
8. Makeup and panning
9. Proofing
10. Baking
11. Cooling
12. Storing

As you can see, mixing ingredients into a dough is only one part of a complex procedure.

Scaling Ingredients

All ingredients must be weighed accurately.

Water, milk, and eggs may be measured by volume. They are scaled at 600gms per1 kg. However, it is more accurate to weigh these liquids, especially if quantities are large.

Special care when measuring spices and other ingredients used in very small quantities. This is particularly important with salt, which affects the rate of fermentation.

Mixing

Mixing yeast doughs has three main purposes:

- To combine all ingredients into a uniform, smooth dough.
- To distribute the yeast evenly throughout the dough.
- To develop the gluten.

Mixing process for doughs and batters in general. This information is important for all dough products and should be reviewed as necessary. The next section contains additional information and procedures needed for the proper mixing of yeast doughs. There are three phases of mixing in the production of doughs and batters. During the first phase, flour and other dry ingredients are hydrated, or combined with liquid ingredients. Bread bakers often refer to this phase as the **pickup stage** because the loose dry ingredients are gradually picked up and incorporated by the developing dough. In the second phase, all the dry ingredients are hydrated and form a rough dough. Bakers call this the **cleanup** stage because the dough forms enough to pull away from (or clean up) the bowl into a compact mass.

In the development phase, the dough is further mixed and gluten is developed to the desired degree. Bakers often divide this phase into two stages, the **initial development** phase, in which the dough still appears rough and undermixed, and **the final development stage**, in which the gluten becomes smooth and elastic. Mixing machines develop doughs by imitating the action of hand kneading. The dough arm or other agitator (depending on the type of mixer) repeatedly stretches the dough and folds it over on itself. This action helps gluten strands not only to form but also to align themselves into a network to make a smooth dough.

Three principal mixing methods are used for yeast doughs: the straight dough method, the modified straight dough method, and the sponge method (also called the sponge-anddough method).

Straight Dough Method

In its simplest form, the straight dough method, also called the direct dough method, consists of only one step: Combine all ingredients in the mixing bowl and mix. With fresh yeast, many bakers make good-quality products using this procedure. However, the yeast may not be evenly distributed in the dough. It is therefore safer to mix the yeast separately with a little of the water. If active dry yeast is used, it is, of course, essential to mix the yeast with water before incorporating it in the dough.

Instant dry yeast, on the other hand, need not be blended with water for use because it is moistened and becomes active in the dough so quickly. The usual method for incorporating instant dry yeast is to mix it in dry form with the flour.

Procedure: Straight Dough Mixing Method

1a.

- In the bowl that is to be used for mixing the dough,soften fresh yeast or active dry yeast in a little of the water.
- Fresh yeast: Mix with about 2 times its weight in water, or more.
- Ideal water temperature: 10 0° F (38°C).Active dry yeast: Mix with about 4 times its weight in water.
- Ideal water temper ature: 105° F (40°C).

1b.

- If using instant dry yeast, mix it directly with the flour.

2. Add the flour to the mixing bowl.
3. Add the remaining ingredients to the top of the flour in the mixing bowl.
4.Mix to a smooth, developed dough.

Modified Straight Dough Method

For rich sweet doughs, the straight dough method is modified to ensure even distribution of the fat and sugar. In this procedure, the fat, sugar, eggs, and flavourings are first blended until uniform before the dough is developed.

Procedure: Modified Straight Dough Method

1. If using fresh or active dry yeast, soften the yeast in part of the liquid, using a separate container. If using instant dry yeast, mix it with the flour.
2. Combine the fat, sugar, salt, milk solids, and flavorings and mix until well combined, but do not whip until light.
3. Add the eggs gradually, but as fast as they are absorbed.
4. Add the liquid; mix briefly. 5. Add the flour and yeast. Mix to a smooth dough.

Sponge Method

Sponge doughs are prepared in two stages. For this reason, the procedure is often called the sponge-and-dough method. This procedure gives the yeast action a head start.

The first stage is called a **sponge**, a **yeast starter**, or a **yeast pre-ferment**. All of these terms can mean the same thing (although bakers often reserve the term sponge specifically for a yeast pre-ferment with a hydration (water content) of about 60 to 63%). Pre-ferments.

There are many variations of the **sponge method**, so the steps indicated in the procedure here are general. Variations on the sponge method an important part in the production of artisan breads. The procedure given here, however, will enable you to prepare the conventional sponge doughs.

A note on the system of baker's percentages is in order here. There are two possible ways to express percentages when using a sponge:

1. Consider the sponge or pre-ferment as a separate formula. Express the flour in the sponge as 100%. Then, in the main formula, express the total weight of the sponge as a percentage of the flour weight in the main formula.
2. Consider the sponge as part of the main formula. Express the flour in the sponge as a percentage of the total flour in the complete formula,

Each method has advantages, and bakers have their own preferences. In this book, both methods are used, depending on the formula, so you can have experience working with each.

Procedure: Sponge Method

1. Combine part or all of the liquid, all of the yeast, and part of the flour (and, sometimes, part of the sugar).

   Mix into a thick batter or soft dough Let ferment until double in bulk
2. Fold (punch down) and add the rest of the flour and the remaining ingredients. Mix to a uniform, smooth dough.

Mixing Times and Speeds

The first two purposes of mixing—combining the ingredients into a dough and distributing the yeast—are accomplished during the first part of this step. The remaining time is necessary to develop the gluten.

To understand mixing times and speeds in detail, it is important to remember that gluten development occurs not only during mixing. It continues during bulk fermentation and folding. Therefore, mixing times, fermentation periods, and number of folds must be balanced. Short mixing time, for example, can be balanced against a long fermentation with numerous folds to yield a properly developed dough. By contrast, if long mixing is followed by long fermentation, the result is an overdeveloped dough.

Three basic mixing techniques are used for most yeast products: the **short mix technique**, the **improved mix** technique, and the **intensive mix** technique. Following the explanation of these techniques:

Short Mix

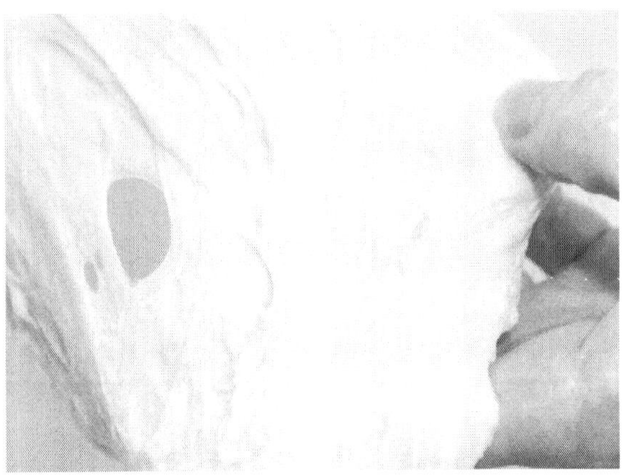

The short mix technique combines a short mix and long bulk fermentation. In a typical planetary mixer, mixing time is 3 to 4 minutes to incorporate ingredients, plus an additional 5 to 6 minutes to develop the gluten. All mixing time is at low speed. Of the three techniques, this is the closest to hand mixing. Because of the short mixing time, the dough is very underdeveloped at the end of the mixing period. Gluten strands have formed but are not well aligned into a smooth dough. As a result, air cells are uneven in size, resulting in an open, irregular crumb structure in the finished bread.

Furthermore, because the dough is underdeveloped after mixing, a long fermentation time is necessary, 3 to 4 hours or even longer. During this long bulk fermentation time, the dough is folded often, generally 4 or 5 times. Because the dough is less developed, it can't hold as much gas and rises less during final proof and baking. Final proof is less than for other mixing techniques.

Shorter mixing also means less oxidation, because less air is mixed into the dough. Less oxidation means less dough strength (a disadvantage). On the other hand, it also means less loss of flavour and colour (an advantage).

The short mix technique is especially suitable for lean doughs with an open crumb, such as baguettes and ciabatta. It is also used for laminated doughs, such croissant dough, in which additional dough development takes place when the butter is rolled in.

Improved Mix

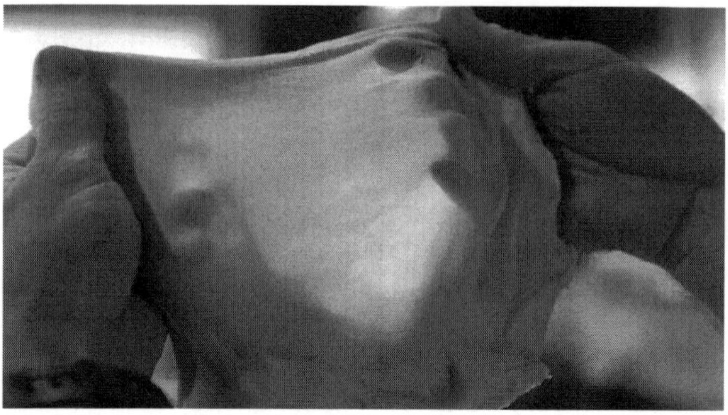

The improved mix technique combines medium mixing and bulk fermentation times. During mixing, the dough is first mixed for 3 to 4 minutes at low speed to incorporate ingredients. Then mixing is continued an additional 5 minutes at medium speed. (Note that the total mixing time is about the same as in the short mix technique, but because second speed is used to develop the gluten, the dough is much more developed than in the short mix.)

Because the dough is more developed after mixing, a shorter fermentation time is needed, usually 1 to 2 hours. Only one or two folds are given during this time.

Improved mix doughs have a slightly whiter colour than short mix doughs, due to oxidation, but they still retain good flavour. Also, the crumb structure is tighter than in short mix products.

This technique is used for many yeast products but is especially appropriate for lean doughs with a slightly open, more regular crumb, and for sweet doughs.

Intensive Mix

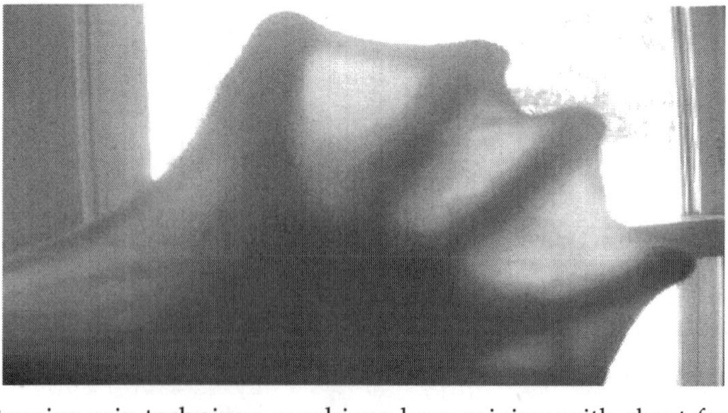

The intensive mix technique combines long mixing with short fermentation. After an initial 3 to 4 minutes at first speed to incorporate ingredients, the dough is then mixed for 8 to 15 minutes at second speed. When the dough leaves the mixer, it has a well-developed gluten structure.

Because the dough is well developed, it can tolerate only a short bulk fermentation, generally about 30 minutes. In most cases, the dough is not folded.

Long mixing means high oxidation, resulting in a very white crumb and only slight aroma and flavour. The crumb is tight and regular.

The intensive mix technique is especially appropriate for nonlaminated doughs high in fat, such as stollen and brioche. When a short production time is important, it can be used for many other breads and rolls, but because of the high level of oxidation, their flavour will be decreased. In Chapter 8 you will find many bread and roll formulas using the intensive mix, because they give you practice with a full range of bread-making procedures in a manageable period of time.

The intensive mix is not appropriate for traditional sourdoughs containing no commercial yeast.

Bulk Fermentation

Fermentation is the process by which yeast acts on the sugars and starches in the dough to produce carbon dioxide gas (CO_2) and alcohol. Gluten development continues during bulk fermentation.

Gluten becomes smoother and more elastic as the dough ferments. For short mix doughs, a fermentation of 3 to 5 hours is required in order to develop the gluten, because it was underdeveloped during mixing. At the opposite extreme, an intensive mix dough needs only about 20 to 30 minutes of bulk fermentation—little more than a bench rest. If it is fermented too long, the dough becomes overdeveloped.

During bulk fermentation, the dough may be folded one or more times, as described in the next section. Folding is an important part of dough development and should be done carefully. In general, the longer the fermentation, the more folds required.

An under fermented dough will not develop proper volume, and the texture of the product will be coarse. An over fermented dough becomes too strong and hard to shape into loaves. If it is fermented far too long, excess acidity forms, making the dough sticky and hard to work. An under fermented dough is called a **young dough**. An over fermented dough is called an **old dough**.

Doughs with weak gluten, such as rye doughs and rich doughs, are usually under fermented, or "taken to the bench young."

At the end of the fermentation time, the dough is mature and ready for dividing and makeup. Through experience, a baker learns to judge dough maturity by feel.

Yeast action continues until the yeast cells are killed, when the temperature of the dough reaches 140°F (60°C) in the oven. It is important to be aware that fermentation continues during the next steps in yeast dough production—folding, dividing, preshaping, benching, and makeup or moulding. Failure to allow for this time may result in over fermented doughs. Doughs that are to be made into rolls and loaves requiring a great deal of makeup time should be slightly under fermented to prevent the dough from being too old by the time makeup is complete

Folding

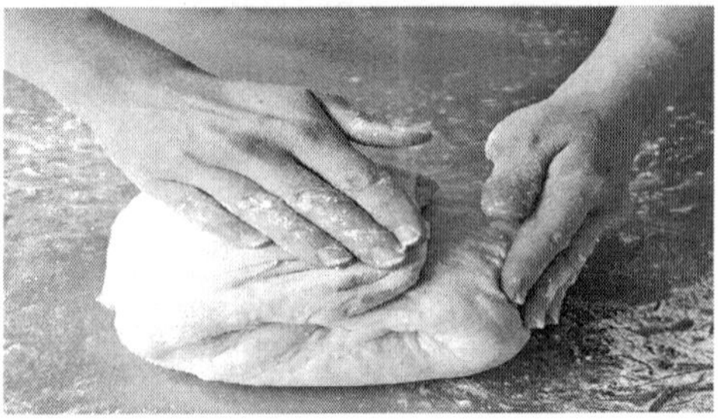

During fermentation, the dough rises, or increases in bulk. When the dough has approximately doubled in bulk, it is folded, as described in the Procedure for Folding Dough, which follows. There are three main benefits of folding:

It expels carbon dioxide. This benefits yeast growth, because yeast activity slows down when its environment contains too much carbon dioxide.

It helps develop gluten structure. The folding process realigns the gluten strands using a technique similar to the action of kneading or mixing.

It equalizes the temperature throughout the dough. Folding should not be done until the dough has expanded in bulk and the gluten is well relaxed and easy to stretch (extensible). If the gluten is still tight, attempting to stretch and fold the dough can harm the gluten structure.

In recipes for home use, folding is usually called **punching**. Some bakers dislike this term, however, because it is misleading, suggesting that the process involves simply hitting the dough with the fist. This is not the case. As just noted, the purpose of folding is not only to deflate the dough but to help develop gluten structure. This cannot be accomplished simply by "punching" the dough.

Large quantities of dough are more easily folded on a workbench, although small batches can easily be folded in the bowl or container in which they are fermented. Both methods are described in the Procedure for Folding Dough.

After folding, if more bulk fermentation is needed, return the dough to the fermentation container. If fermentation is complete, bring the dough to the workbench to prepare for the next step: dividing.

Dividing (Scaling or Portioning Dough)

Using a baker's scale, divide the dough into pieces of the same weight, according to the product being made.

In other words, there are two phases to this stage of bread production:

1. Cutting the dough into pieces.
2. Scaling the pieces to check and adjust their weight.

Try to cut the dough into pieces as close as possible to the desired weight. After cutting and scaling the first few pieces, you should be able to estimate the correct size to within an ounce or two. When each piece is scaled, adjust the weight as needed by adding or removing a small piece of dough.

During scaling, allowance is made for weight loss due to evaporation of moisture in the oven. This weight loss is approximately 10 to 13% of the weight of the dough. Allow an extra 50 to 65 grams per 500 grams. Actual baking loss depends on baking time, size of the unit, and whether it is baked in a pan or freestanding. If a dough divider is used to make rolls, the dough is scaled into presses, which are then divided into 36 equal pieces. Presses are rounded, relaxed, and divided; the divided units may or may not be rounded again, depending on the product.

Rounding

After scaling, the pieces of dough are preshaped. The goal is to form the dough into a fairly smooth, regular piece that, after resting, can easily be made into the final shape.

Many bakers preshape all their dough into smooth, round balls. This process is called **rounding**. Rounding forms, a kind of skin by stretching the gluten on the outside of the dough into a smooth layer. This skin gives the dough a uniform shape and helps retain gases produced by the yeast.

Although it is possible to shape nearly any kind of loaf from a rounded dough, many bakers prefer to preshape the dough pieces into forms that are closer to their final shape. Thus, for long loaves such as baguettes, the dough is preshaped into a cylinder. Final shaping then requires less handling of the dough.

As in fermentation, folding, and dividing, handle the dough during shaping so as to not damage its structure. In the case of breads that have an open crumb structure, do not compress the dough too much. Handle it gently. On the other hand, for breads that need a tight crumb, such as Pullman loaves and bagels, handle the dough more vigorously to eliminate any large holes.

Also, evaluate the strength of the dough before preshaping and handle it accordingly. If the dough is very extensible (stretchable), give it a tighter preshaping. If it resists stretching, give it a loose preshaping so that the gluten is not too strong for final makeup.

Bench-Proofing or Intermediate Proofing

Rounded or preshaped portions of dough are allowed to rest for 10 to 20 minutes.

This relaxes the gluten to make final shaping of the dough easier. Also, fermentation continues during this time.

In large operations, the rounded dough is placed in special proofers for this rest. Smaller operations place the dough in boxes that are stacked on one another to keep the dough covered. Or the dough may simply be placed on the workbench and covered—hence the term benching or bench rest.

Makeup and Panning

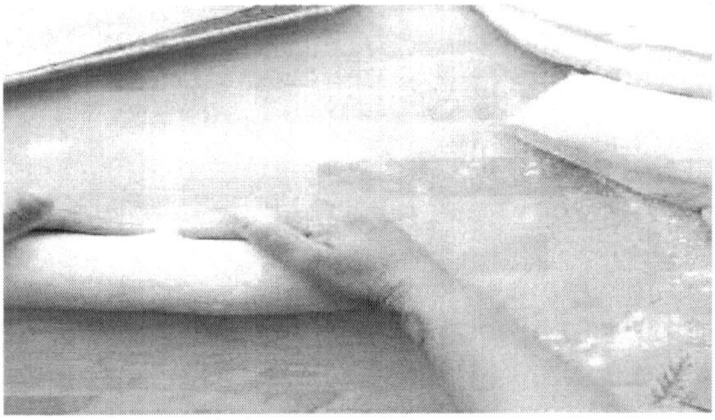

The dough is shaped into loaves or rolls and then placed in pans or on baking sheets. **Hearth breads**—breads baked directly on the bottom of the oven—may be placed in floured baskets or other moulds after makeup. Proper makeup or moulding is of critical importance to the finished baked product. If a tight crumb is desired, all gas bubbles should be expelled during molding. Bubbles left in the dough will result in large air holes in the baked product. By contrast, breads with an open crumb should be molded more gently to preserve the air holes.

For both pan breads and hearth breads, the seam must be centered on the bottom to avoid splitting during baking. For units baked in pans, the pan size must be matched to the weight of the dough. Too little or too much dough will result in a poorly shaped loaf.

Proofing

Proofing is a continuation of the process of yeast fermentation that increases the volume of the shaped dough. Bakers use two terms so they can distinguish between fermentation of the mixed dough and proofing of the made-up product before baking. Proofing temperatures are generally higher than fermentation temperatures.

Under proofing results in poor volume and dense texture. Over proofing results in coarse texture and some loss of flavour.

For short mix doughs, final proofing time is relatively short—up to 1 hour, on average— because the gluten is not as well developed and therefore will not hold as much gas. Proofing for improved and intensive mix doughs generally takes 1 to 2 hours. For breads leavened only by sourdough starter and no commercial yeast, proofing takes even longer.

Rich doughs are slightly under proofed because their weaker gluten structure does not withstand too much stretching

Baking

Many changes take place in the dough during baking. The most important changes are:

- Oven spring, also called oven kick, which is the rapid rising in the oven due to production and expansion of trapped gases as a result of the oven heat. The yeast is very active at first but is killed when the temperature inside the dough reaches 140°F (60°C).
- #Coagulation of proteins and gelatinization of starches. In other words, the product becomes firm and holds its shape.
- Formation and browning of the crust.

To control the baking process, consider the factors described next.

Oven Temperature and Baking Time

Temperatures must be adjusted for the product being baked. At the proper temperature, the inside of the unit becomes completely baked at the same time the crust achieves the desired colour. Therefore:

1. Large units are baked at a lower temperature and for a longer time than small rolls spaced apart.
2. Rich doughs and sweet doughs are baked at a lower temperature because their fat, sugar, and milk content makes them brown faster.
3. French breads made with no added sugar and a long fermentation require very high temperatures to achieve the desired crust colour.

   Note the following specific temperatures:

   - Popular American lean breads are baked at 400° to 425°F (205° to 220°C).
   - Some French breads are baked at 425° to 475°F (220° to 245°C).
   - Rich products are baked at 350° to 400°F (175° to 205°C).
   - A golden-brown crust colour is the normal indicator of doneness. Loaves that are done sound hollow when thumped.

Washes

Many, if not most, yeast products are brushed with a liquid, called a wash, just before baking. The most common washes are as follows:

Water is used primarily for hard-crusted products, such as French bread. Like steam in the oven (see below), the water helps keep the crust from drying too quickly and thus becoming too thick.

Starch paste is used primarily for rye breads. In addition to keeping the crust from drying too quickly, the starch paste helps give a shine to the crust. To make a starch paste, mix 30g light rye flour with60 g rye per 500 mL water. Bring to a boil while stirring. Cool. If necessary, thin with water to the consistency of cream.

Egg wash is used to give a shiny brown crust to soft breads and rolls and to rich doughs and Danish. It is made by mixing beaten eggs with water or, sometimes, milk. Proportions vary greatly depending on how strong a wash is desired.

Commercial aerosol washes (sprays) provide a quick and easy way to give shine and to help toppings such as seeds adhere.

Scoring

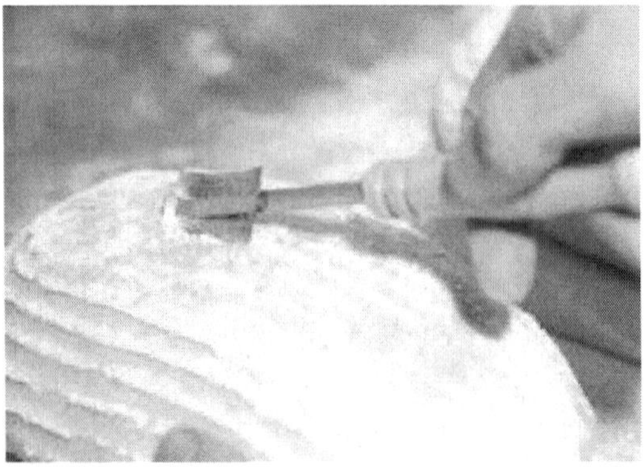

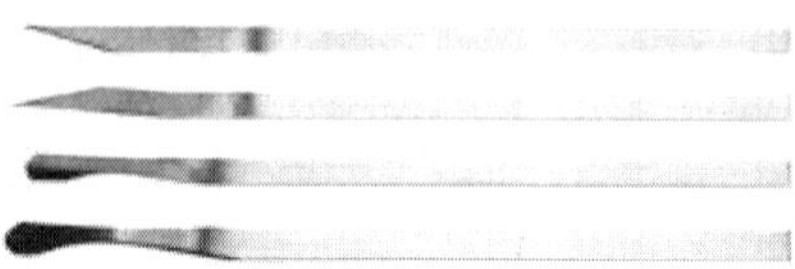

Baker's lames.

A break on the side of the loaf is caused by continued rising after the crust is formed. To allow for this expansion, the tops of hard-crusted breads are cut just before baking. Properly scored loaves expand better in the oven and attain greater volume and have a more open crumb structure.

Slashes are made on the top of the loaf with a baker's lame (pronounced lahm; French for "blade," a curved or straight razor blade attached to a handle) or another sharp knife or razor immediately before it is put into the oven, as shown in the photograph. The pattern created by the cuts also contributes to the appearance of the bread. Follow these scoring guidelines:

- If the loaves seem somewhat over proofed, or if the gluten is weak, make only shallow cuts to prevent deflating the loaves. Under proofed loaves can be scored more deeply.
- Small rolls usually bake completely without a break, so they are usually not scored unless desired for the sake of appearance.
- Loaves baked in pans are usually not scored, because the pan enables the bread to expand without rupturing.

Loading the Ovens

Proofed doughs are fragile until they become set by baking. Therefore, handle them carefully when loading them into the ovens, and do not disturb them during the first part of baking.

Breads and rolls are baked either directly on the bottom of the oven (hearth breads) or in pans.

- **Hearth breads.** To load the ovens, place the proofed units on a peel that is well dusted with cornmeal. Slide the peel into the oven. Then, with a quick snap, remove the peel, leaving the loaves or rolls in place. To remove baked items, quickly slide the peel under them and pull them out. The illustration shows a peel in use.
- **Pan breads and rolls.** Freestanding items may be baked on sheet pans instead of on the hearth. Bakers generally refer to such breads and rolls as hearth breads even if they are not baked directly on the bottom of the oven. Sprinkle the pans with cornmeal to keep the units from sticking, and to simulate the appearance of hearth-baked items. Pans may also be lined with silicone paper. Perforated sheet pans or screens are also available. These allow better air circulation and therefore permit more even browning.

Sandwich loaves and other pan breads are, of course, baked in loaf pans or other appropriate pans.

Cooling

After baking, bread must be removed from the pans and cooled on racks to allow the escape of the excess moisture and alcohol created during fermentation.

Small rolls spaced on baking sheets are often cooled on the pans when air circulation is adequate. On the other hand, if condensation is likely to make the bottoms of the rolls soggy, it is better to cool them on racks.

If soft crusts are desired, breads may be brushed with melted shortening before cooling.

Do not cool bread in a draft because the crust may crack.

Like other dough products, breads continue to undergo physical and chemical changes after they are removed from the oven. Review page 102 for a summary of these changes.

Storing

Breads to be served within eight hours may be left on racks. For longer storage, wrap cooled breads in moisture proof bags to retard staling. Note, however, that bread must be thoroughly cool before wrapping, or moisture will collect inside the bags.

Wrapping and freezing maintains quality for longer periods. Refrigeration, on the other hand, increases staling.

Hard-crusted breads should not be wrapped (unless frozen) because the crusts will soften and become leather y. Alternatively, use porous bags or wrapping material that protects the bread from contamination but allows moisture to escape.

Bread Mixing Methods

There are 5 methods of bread-making

1. Straight dough method.
2. No-time dough method
3. Delayed salt method.
4. Sponge and dough method.
5. Ferment and dough method.

Straight Dough Method

In this method all the ingredients are mixed together, and the dough is fermented for a predetermined time.

The fermentation time depends on the strength of the flour, strong flours require longer fermentation time to mature adequately.

Flours which require 2 -3 hours for maturing should be used for making bread by straight method. Flours that take very long for maturing should not be used for straight dough methods because during prolonged fermentation it is difficult to control the temperature of the dough and rise in temperature causes acid taste and flavour in bread.

No-time Dough Method

In this method dough is not fermented in the ususal manner. It is allowed to ferment for a short period so the twin function of fermentation ie. production of gas and conditioning of gluten are achieved to some extent by increasing the amount of yeast.

And by making the dough a little slack and warm.

It is possible to get a good product using this method, but the product has poor keeping quality and lacks aroma. Due to short fermentation time, the gluten and starch are not conditioned to hold moisture and there is no flavour because flavour producing bi- products of fermentation are absent. As there is increased quantity of yeast present, the bread may have a strong yeast flavour.

Delayed-Salt Method

This is a slight variation of the straight method, where all the ingredients are mixed except salt and fat.As salt has a controlling action on the yeast function, the speed of fermentation of a saltless dough will be faster and a reduction in the fermentation time is affected.The salt is added at the knock-back stage.The method of addingd salt depends on the convenience of indiviual bakers.

It may be sifted on the dry dough or it is creamed with fat and incorporated into the dough.

This method is suitable for strong flours if straight dough method is used.

Sponge and Dough Method

This method is normally used for strong flours. Sponge doughs are prepared in 2 stages. This procedure gives yeast action a head start.

The 1st stage is called a sponge, a yeast starter or a yeast ferment. All mean the same thing.

In this method, a part of flour, proportionate amount of water, all the yeast and yeast food are mixed together. Mixing operation are carried out to incorporate all the ingredients evenly. The sponge is fermented fora pre-determined time, which on the quality of flour, and the amount of flour in the sponge.

The sponge is then physically tested, by either taking a piece of sponge and try to break it with both hands,. If the piece breaks with a clean fracture then the sponge is ready for mixing If the sponge stretches and breaks unevenly, then some more fermentation time is needed.

Tear the sponge apart with both hands and examine the web structure, if the web structure is very fine then the sponge is ready.

An adequately fermented dough feels dry to touch without any stickiness present.

When the sponge is ready, it should be broken down properly with the formula water, and mixed with the remaining flour, sugar, salt and fat. After the dough is mixed, it is rested for 30-40 minutes'-conditioning of the gluten [during the sponge stage] hastens the conditioning process.

For the sake of identification and convenience, a sponge is indicated as 60/40 or 70/30, where the first number indicates the percentage of flour used in the sponge.

Ferment and Dough Method

This a variation of the sponge and dough method. Rich doughs which contain milk, eggs, substantial amounts of fat and sugar, have a retarding effect on yeast activity. If all the formula yeast, part of the flour. yeast food and sufficient water is mixed together, the yeast get initially an environment which is conducive to vigorous activity and it is in a fit condition to take on the extra load of fermentation in the presence of milk, egg etc.

When the ferment is ready, it is mixed into the dough, along with the remaining ingredients, along with the remaining ingredients and allowed to ferment [second stage]. This method is used in making enriched breads, buns, Danish pastry etc.

Characteristics of Good Bread

To define a good loaf, one must have some knowledge of the desirable qualities of a particular type of bread and how these qualities are produced.

The desirable features of a good loaf can be listed under two headings, external and internal.

External	Internal
Volume	Grains
Colour of crust	Texture
Evenness of bake	Colour of crumb
Symmetry of form	Aroma
Break and shred	Taste
Character of crust	

External Characteristics

Volume

- Volume is an important consideration in consumer acceptance.
- The greater volume of bread, the softer a loaf appears when squeezed with hands.
- A loaf of excessive volume will have generally have open grain and weak texture while the one with a low volume exhibits rough grain and honeycomb texture.

Colour of Crust

- The color of crust is also termed as "bloom".
- The desired color is one with an appetizing golden-brown shade.
- Crust color is the result of caramelisation of sugars and due to the browning reaction due to interaction of protein and reducing sugar.
- The color of crust is mainly dependent on the amount of sugar present in the dough when baked and temperature at which the loaf is baked.

Evenness of Bake

- This means that the loaf has been made with a uniformity of bake on all sides.
- Factors like correct Proofing, controlled oven conditions as well as the proper distance between the pans during baking affect evenness of baking.

Symmtery of Form

The shape or symmetry of the loaf influences the customers to the great extent.

Break and Shred

- The break and shred of a loaf have an important bearing on the general appearance of the loaf.
- An even- shredded break is desirable on the sides and ends of open top bread.
- The break should not be excessive to the extent that the top crust separates from the sides of the loaf forming crust.
- Faulty bread may show any of these faults wild bread, lack of attractive shred, shell top, no break or insufficient break.

Character of Crust

- The bread crust should be thin and crisp. It should not be thick, tough and rubbery.

Internal Characteristics

- The most important characteristics of a loaf of a bread is crumb structure.
- This is broadly divided into grains and texture.

Grains

- Cell structure of a loaf is known as grain.
- A fine silky grain is one in which cells are small and uniform, elongated, rather than round and cell walls are thin.
- A coarse grain is one in which the cells are large and the cells wall are thick.

Texture

- Texture is actually the feel of the surface of the interior of the loaf when cut and sliced.
- Thus, a desirable texture may be classifies as velvety, silky, soft and elastic.
- An inferior texture is often described as rough, hard, doughy, crumbly and lumpy.
- The texture of a loaf is determined by pressing the fingers against and rubbing them across the cut surface of the loaf.

Colour of Crumb

- A coarse grain slice will appear darker in color and fine grain slice whiter even though both loaves were made with identical ingredients.
- Thick heavy cell walls due to young dough give the crumb a little tan color.
- Open grain causes shadows which give the crumb a grayish color.
- The color of the crumb should be bright without any dark patches.

Aroma

- The aroma of a loaf is determined by carefully smelling it.
- The aroma may be described as wheaty, sweet, musty, rancid, moldy, sour or flat.

- The ideal loaf has a pleasant wheaty and fermentation aroma.

Taste

- The aroma and taste of a loaf are closely related characteristics.
- Taste is actually the taste of the loaf when eaten.
- It can be described as wheaty, sweet, sour, flat or rancid.

Bread Faults and their Remedies

A thorough knowledge about raw material and its functions, adequate understanding of bread making procedure, control of temperature and humidity at different stages of bread making and above all personal skill and experience od baker goes a long way in avoiding faults in bread. There are number of factors which are responsible for creating faults in bread.

Major factors which adversely influence the quality of bread are:

- Inadequate gluten in flour
- Misappropriate quantities and inferior quality of raw material
- Poor diastatic activity of flour
- Improper time and temperature of fermentation, proofing and baking
- Wrong methods of manipulation of dough I.e. knocking-back, cutting and moulding
- Inadequate cooling of bread
- Improper storage of bread and
- Lack of knowledge about the principles of hygiene.

The following are some of the major faults in bread:

- Volume: Volume of the bread is the outcome of adequate conditioning of gluten and sufficient gassing power of the dough at the time of baking.
- Small Volume of Bread may Be Due to Excess Volume Can Be Due to

Small Volume of Bread may Be Due to	Excess Volume Can Be Due to
• Tight dough	• Over fermentation
• Little yeast and fermentation time	• Lack of salt in formula
• Low temperature	• Excessive yeast and proofing time
• Under proofing	• Loose moulding
• Lack of diastatic activity	• Lack of temperature in oven or cool oven
• Bran contamination	• If yeast is added in excess, it will consume more of sugar and bread will be light and pale brown.
• Under mixing or over mixing	• Insufficient temperature will cause lack of crust colour
• Very high temperature during baking	• Insufficient humidity during proofing
• Too long intermediate proof	• Under baking
• Addition of excess of salt	• Oven temperature is low
• Use of weak flour	• Poor diastatic activity of flour
• Use of less amount of shortening	• Old dough

FAULTS	CAUSES
SHAPE	Too much salt
Poor volume	Too little yeast
	Too little liquid
	Weak flour
	Under or overmixing
	Oven too hot
Too much volume	Too little salt
	Too much yeast
	Too much dough scaled
	Over proofed
Poor shape	Too much liquid
	Flour too weak
	Improper molding or makeup
	Improper fermentation of proofing
	Too much oven steam
Split or burnt crust	Overmixing
	Underfermented dough
	Improper molding-seam not on bottom
FLAVOR	
Flat taste	
Poor flavor	Too little salt
TEXTURE AND CRUMB	
Too dense or closed-grain	
	Inferior, spoiled, or rancid ingredients
	Poor bakeshop sanitation
	Under-or overfermented
Too coarse or open	Too much salt
	Too little liquid
	Too little yeast
	Underfermented
	Underproofed
	Too much yeast
Streaked crumb	Too much liquid
	Incorrect mixing time

	Improper fermentation
	Overproofed
	Pan too large
Poor texture or crumbly	Improper mixing procedure
	Poor molding or makeup techniques
	Soo much flour used for dusting
	Flour too weak
	Too little salt
Gray crumb	Fermentation time too long or too short
	Overproofed
	Baking temperature too low
CRUST Too dark	Fermentation time or temperature too high
	Too much sugar or milk
	Underfermented dough
Too pale	Oven temperature too high
	Baking time too long
	Insufficient steam at beginning of baking
	Too little sugar or milk
	Overfermented dough
Too thick	Overproofed
	Oven temperature too low
	Baking time too short
	Too much steam in oven
	Too little sugar or fat
	Improper fermentation
Blisters on crust	Baked too long or at wrong temperature
	Too little steam
	Too much liquid
	Improper fermentation
	Improper shaping of loaf

Understanding of Bread Faults

No oven spring

Solution

(a) Check recipe
(b) Decrease dough temperature
(c) Reduce resting time
(d) Increase mixing time.

Blisters and holes on crust

Solution

(a) Check recipe
(b) Increase resting time
(c) Mould correctly
(d) Check proper humidity
(e) Reduce oven steam

Excessive colour

Solution

(a) check recipe
(b) increase dough temperature
(c) increase resting time
(d) reduce baking temperature

Coarse crumb structure

Solution

(a) reduce water level
(b) increase dough temperature
(c) increase mixing time
(d) increase resting time
(e) reduce proof time and temperature

Rapid staling

Solution

(a) check recipe
(b) increase water level
(c) reduce resting time
(d) increase baking temperature
(e) reduce baking time

Collapsed rolls or fancy breads

Solution

(a) check recipe
(b) increase mixing time
(c) reduce proof time and temperature

Poor volume

Solution

(a) check recipe
(b) Increase water level
(c) Increase mixing time
(d) Increase resting time
(e) Check proofing time
(f) Reduce baking time

BREAD FAULTS IN BREAD LOAF

Flying tops

Solution

(a) Rest dough for longer period
(b) Extend final proof time
(c) Decrease initial baking temperature
(d) Use steam while baking

Collapsed bread

Solution

(a) Check flour quantity
(b) Check recipe
(c) Reduce water level
(d) Increase mixing time
(e) Ensure correct final proof

Badly shaped bread

Solution

(a) Increase water level
(b) Check moulder settings
(c) Mould dough evenly
(d) Check dough placement on pan
(e) Use steam when baking

Wrinkled crust

Solution

(a) Check recipe
(b) Extend resting time
(c) Reduce final proof time
(d) Cool bread before wrapping
(e) Ensure packed at correct temperature

Holes under top crust

Solution

(a) Check recipe
(b) Reduce mixing time
(c) Reduce resting time
(d) Reduce final proof time
(e) Reduce oven top heat

Uneven texture

Solution

(a) Check water level
(b) Increase resting time
(c) Mould dough evenly
(d) Check proof height

Rapid staling

Solution

(a) Check recipe
(b) Increase water level
(c) Reduce resting time
(d) Increase baking temperature
(e) Reduce baking time

Mould and rope

Solution

(a) Ensure hygiene and cleanliness
(b) Check recipe preservative level
(c) Check baking time
(d) Ensure sliced and packed at correct temperature
(e) Ensure correct storage environment

Sides collapsing

Solution

(a) Check recipe
(b) Reduce mixing time
(c) Extend baking time
(d) Remove from pans after baking

Bread sticks in pan

Solution

(a) Reduce water level
(b) Check pan oil application
(c) Check proofer steam setting
(d) Reduce final proof time
(e) Extend baking time

5. COOKIES AND BISCUITS

Cookies

The word cookie means "small cake," and that's more or less what a cookie is. in fact, some cookies are made from cake batter. for some products, such as certain kinds of brownies, it's difficult to know whether to classify them as cakes or cookies.

Most cookie formulas, however, call for less liquid than cake formulas do. Cookie doughs range from soft to very stiff, unlike the thinner batters for cakes. This difference in moisture content means some differences in mixing methods, although the basic procedures are much like those for cakes.

The most obvious differences between cakes and cookies are in makeup. Because most cookies are individually formed or shaped, a great deal of hand labour is involved. Learning the correct methods and then practicing diligently are essential for efficiency.

Cookies come in an infinite variety of shapes, sizes, flavors, and textures. Characteristics that are desirable in some types are not desirable in others. For example, we want some cookies to be crisp, others to be soft. We want some to hold their shape, others to spread during baking. In order to produce the characteristics, we want, and to correct faults, it is useful to know what causes these basic traits.

- Crispness
- Softness
- Chewiness
- Spread

Mixing Methods

There are four basic cookie mixing methods:

1. One-stage
2. Creaming
3. Sponge

1. One-Stage Method

The one-stage method is the counterpart of the one-stage cake-mixing method. As just noted, cookie dough contain less liquid than cake, batters do, so blending the ingredients into uniform dough is easier. Because all the ingredients are mixed at once, the baker has less control over mixing with this method than with other methods. Therefore, the one-stage method is not frequently used. When over mixing is not a great problem, as with some chewy cookies, it can be used.

Procedure

1. Scale ingredients accurately. Have all ingredients at room temperature.
2. Place all ingredients in the mixer. With the paddle attachment, mix the ingredients at low speed until uniformly blended. Scrape down the sides of the bowl as necessary.

2. Creaming Method

The creaming method for cookies is nearly identical to the creaming method for cakes. Because cookies require less liquid than cakes, it is not usually necessary to add the liquid alternately with the flour. It can be added all at once.

Note the importance of step 2 of the procedure, the creaming stage. The amount of creaming affects the texture of the cookie, the leavening, and the spread. Only a small amount of creaming is desired when the cookie must retain its shape and not spread too much. Also, if the cookie is very short (high in fat and low in gluten development), or if it is thin and delicate, too much creaming will make the cookie too crumbly.

Procedure

1. Scale ingredients accurately. Have all ingredients at room temperature.
2. Place the fat, sugar, salt, and spices in the mixing bowl. With the paddle attachment, cream these ingredients at low speed. Partway through mixing, stop the machine and scrape down the bowl to ensure even mixing.
3. For light cookies, cream until the mix is light and fluffy, in order to incorporate more air for leavening. For denser cookies, blend to a smooth paste, but do not cream until light.
4. Add the eggs and liquid, if any, and blend in at low speed.
5. Sift in the flour and leavening. Mix until just combined. Do not overmix, or gluten will develop.

Sponge Method

The sponge method for cookies is similar to the egg-foam methods for cakes. The procedure varies considerably, however, depending on the ingredients. Batches should be kept small because the batter is delicate.

Procedure

1. Scale all ingredients accurately. Have all ingredients at room temperature except the eggs, which you may want to warm slightly for greater volume, as for sponge cakes.
2. Following the procedure given in the formula being used, whip the eggs (whole, yolks, or whites) and the sugar to the proper stage: soft peaks for whites, thick and light for whole eggs or yolks.
3. Fold in the remaining ingredients as specified in the recipe. Be careful not to overmix or to deflate the eggs.

Types and Makeup Methods

We can classify cookies by their makeup methods as well as by their mixing methods. Grouping them by makeup method is perhaps more useful, from the point of view of production, because cookie mixing methods are relatively simple, while their makeup procedures vary considerably.

- Bagged
- Dropped
- Rolled
- Molded
- Icebox
- Bar
- Sheet
- Stencil

Panning, Baking, and Cooling

Preparing the Pans

1. Use clean, un warped pans.
2. Lining the sheets with parchment or silicone paper is fast, and it eliminates the need to grease the pans.
3. A heavily greased pan increases the spread of the cookie. A greased and floured pan decreases spread.
4. Some high-fat cookies can be baked on ungreased pans.

Baking

1. Most cookies are baked at a relatively high temperature for a short time.
2. Too low a temperature increases spreading and may produce hard, dry, pale cookies.

Preparing the Pans

1. Use clean, un warped pans.

2. Lining the sheets with parchment or silicone paper is fast, and it eliminates the need to grease the pans.
3. A heavily greased pan increases the spread of the cookie. A greased and floured pan decreases spread.
4. Some high-fat cookies can be baked on ungreased pans.

Baking

1. Most cookies are baked at a relatively high temperature for a short time.
2. Too low a temperature increases spreading and may produce hard, dry, pale cookies.

Cooling

1. Most cookies baked without parchment paper must be removed from the pans while they are still warm, or they may stick.
2. If the cookies are very soft, do not remove them from the pans until they are cool enough and firm enough to handle. Some cookies are soft when hot but become crisp when cool.
3. Do not cool cookies too rapidly or in cold drafts, or they may crack.
4. Cool completely before storing. After the cookies have been baked, check them for defects.

Cookie Faults and their Causes

Fault	Causes
Too tough	Flour too strong
	Too much flour
	Not enough shortening
	Incorrect amount of sugar
	Mixed too long or improper mixing
Too crumbly	Improper mixing
	Too much sugar
	Too much shortening
	Too much leavening
	Not enough eggs
Too hard	Baked too long or,
	baking temperature too low
	Too much flour
	Flour too strong
	Not enough shortening
	Not enough liquid

Fault	Causes
Too dry	Not enough liquid
	Not enough shortening
	Baked too long or,
	baking temperature too low
	Too much flour
Not browned enough	Baking temperature too low
	Underbaked
	Not enough sugar
	Too brown Baking temperature too high
	Baked too long
	Too much sugar
Poor flavor	Poor-quality ingredients
	Flavoring ingredients left out
	Dirty baking pans
	Ingredients improperly measured
	Sugary surface or crust Improper mixing
	Too much sugar
Too much spread	Baking temperature too low
	Not enough flour
	Too much sugar
	Too much leavening,
	(chemical leaveners or creaming)
	Too much liquid
	Pans greased too heavily
Not enough spread	Baking temperature too high
	Too much flour or flour too strong
	Not enough sugar
	Not enough leavening
	Not enough liquid
	Insufficient pan grease
Stick to pans	Pans improperly greased
	Too much sugar
	Improper mixing

Chapter 5

Confectionery

1. Cake

Ingredients Used in Cake Making

Cake Flour

- Cake flour is a weak or low-gluten flour made from soft wheat.
- It has a soft, smooth texture and a pure white color.
- Cake flour is used for cakes and other delicate baked goods that require low gluten content.
- Protein content of cake flour is approximately 8%, and ash content is approximately 0.3%.

Sugar

- When a sucrose solution is heated with an acid, some of the sucrose breaks down into equal parts of two simple sugars, dextrose and levulose.
- A mixture of equal parts of dextrose and levulose is called invert sugar.
- It is about 30% sweeter than regular sucrose.
- Invert sugar has two properties that make it interesting to the baker.
- First, it holds moisture especially well—that is, it is very hygroscopic—and, therefore, helps keep cakes fresh and moist.
- Second, it resists crystallization.
- Thus, it promotes smoothness in candies, icings, and syrups.
- Invert sugar is produced commercially and is available as a syrup.
- It is also present in honey.

Shortenings

- Any fat acts as a shortening in baking because it shortens gluten strands and tenderizes the product.
- However, we generally use the word shortening to mean any of a group of solid fats, usually white and tasteless, that are especially formulated for baking.
- Shortenings generally consist of nearly 100% fat.
- This process turns liquid oils into solid fats.
- Shortenings are used for many purposes, so manufacturers have formulated different kinds of fats with various properties.
- There are three main types: regular or all-purpose (AP) shortenings, high-ratio plastic shortenings, and high-ratio liquid shortenings.

Eggs

- Eggs should be well understood by the baker because they are used in large quantities in the bakeshop and are more expensive than many of the other high-volume ingredients, such as flour and sugar.
- For example, half or more of the ingredient cost of the average cake batter is for the eggs.

Moistening Agent

- Liquids have several functions in baking.
- Liquids are necessary in baked foods for hydrating protein, starch and leavening agents.
- When hydration occurs, water is absorbed and the chemical changes necessary for structure and texture development can take place.
- Liquids induce enzyme activity, help in dispersing non flour ingredients uniformly and dissolve salts.
- They also contribute moistness to the texture and improve the mouth feel of baked products.
- When water vaporizes in a batter or dough, the steam expands the air cells, increasing the final volume of the product.
- Milk contributes water and valuable nutrients to baked foods. It helps browning to occur and adds flavor.
- Liquids keep bread palatable longer if they are retained in the finished loaf.
- When making yeast dough, milk should be scalded and cooled before adding to other ingredients.
- This is done to improve the quality of the dough and the volume of the bread and should not be substituted with water.
- Fruit juices may be used as the liquid in a recipe if water is specified.
- Because fruit juices are acidic, they are probably best used in baked products that have baking soda as an ingredient.
- But juice in bread baking slows fermentation because of its high sugar which reduces yeasts activity.

Leavening Agents

- Unlike liquids and solids, gases expand greatly when they are heated.
- Leavening is the production or incorporation of gases in a baked product to increase volume and produce shape and texture.
- These gases must be retained in the product until the structure is set enough to hold its shape.
- The three main gases that leaven baked goods are carbon dioxide, steam, and air.
- Two of these gases, steam and air, are present in all baked goods.
- Leavening gases are trapped in these air cells, which expand as they fill with gases and as the gases are heated.
- The walls of the cells are formed largely from gluten proteins and sometimes egg proteins.
- They hold the gases and form the structure of the baked item.

- Exact measurement of leavening agents is important because minor changes can produce major defects in baked products.

Cake Making Methods

Sugar Butter Process

1. Place the butter out on the counter for at least an hour, or until it becomes room temperature. (This gives you time to measure out the rest of the ingredients in your recipe and preheat your oven.) The butter should feel soft, but not warm or melty.
2. Slice the butter into cubes or grate using the largest side of a grater, and put the cubes/shreds in a large mixing bowl. Beat the butter with a wooden spoon until it is soft.
3. Add your sugar(s) to the butter and gently mash it into the butter with the tines of a fork. With your wooden spoon, stir the butter and sugar(s) until they are light and fluffy. Use a rubber spatula to scrape the mixture off the sides of the bowl periodically. The butter is "creamed" when it has almost doubled in mass and it has lightened to a yellowish-white color.

Flour-Batter Method

1. Scale all ingredients accurately. Have all ingredients at room temperature.
2. Sift flour & other dry ingredients except sugar into mixing bowl. Add fat. Blend until smooth & light.
3. Whip the sugar and eggs together until thick and light. Add liquid flavoring ingredients, such as vanilla.
4. Combine the flour-fat mixture and the sugar-egg mixture and mix until smooth.
5. Gradually add water or milk (if any) and mix smooth.

Genoise Method

1. Scale all ingredients accurately.
2. Combine the eggs, sugar, and salt in a stainless steel bowl. Immediately set the bowl over a hot-water bath and stir or beat with a whip until the mixture warms to a temperature of about 110°F(43°C). The reason for this step is that the foam attains greater volume if warm.
3. With a wire whip or the whip attachment of a mixer, beat the eggs at high speed until they are very light and thick. This may take as long as 10 to 15minutes if the quantity is large.
4. If any liquid (water, milk, liquid flavoring) is included, add it now. Either whip it in, in a steady stream, or stir it in, as indicated in the recipe.
5. Fold in the sifted flour in 3 or 4 stages, being careful not to deflate the foam. Many bakers do this by hand, even for large batches. Fold gently until all the flour is blended in. If any other dry ingredients are used, such as cornstarch or baking powder, sift them first with the flour.
6. Immediately pan and bake the batter. Delays will cause loss of volume.

Blending and Rubbing Method

1. Use a freezing cold fat. When using the rubbing-in method to make scones and pastries the fat and liquid needs to be cold. If using butter, freeze it and then grate it into the flour before rubbing it in. Because when cold butter is rubbed into the flour, it creates flaky pockets of flavor (which soft, room temperature butter cannot do). Once the cold butter and liquid (e.g. milk) hits the oven, the water in the butter and cold liquid begins evaporating. This creates moisture, and helps the dough to rise, creating a tender and light crumb.
2. Use your fingertips only when rubbing the butter in to the flour/dry mixture, try not to use your whole hand to do so. The palm of your hand is warmer than your fingertips and will soften the butter. If it is a hot day, you might be better off getting out the food processor to do the job for you. Aerate the mixture. Try keep your fingertips above the mixture when rubbing the butter in to create more air.

Basic Pastries

Pastry is a dough of flour, water and shortening (solid fats, including butter) that may be savoury or sweetened. Sweetened pastries are often described as *bakers' confectionary*. The word "pastries" suggests many kinds of baked products made from ingredients such as flour, sugar, milk, butter, shortening, baking powder, and eggs. Small tarts and other sweet baked products are called **pastries**. Common pastry dishes include pies, tarts, quiches, croissants, and pasties.

The French word patisserie is also used in English (with or without the accent) for the same foods. Originally, the French word *patisserie* referred to anything, such as a meat pie, made in dough (*paste,* later *pâte*) and not typically a luxurious or sweet product. This meaning still persisted in the nineteenth century, though by then the term more often referred to the sweet and often ornate confections implied today.

Pastry can also refer to the **pastry dough.** from which such baked products are made. Pastry dough is rolled out thinly and used as a base for baked products.

Pastry is differentiated from bread by having a higher fat content, which contributes to a flaky or crumbly texture. A good pastry is light and airy and fatty, but firm enough to support the weight of the filling. When making a shortcrust pastry, care must be taken to blend the fat and flour thoroughly before adding any liquid. This ensures that the flour granules are adequately coated with fat and less likely to develop gluten. On the other hand, overmixing results in long gluten strands that toughen the pastry. In other types of pastry such as Danish pastry and croissants, the characteristic flaky texture is achieved by repeatedly rolling out a dough similar to that for yeast bread, spreading it with butter, and folding it to produce many thin layers.

Types

Short Crust Pastry

Shortcrust pastry is the simplest and most common pastry. It is made with flour, fat, butter, salt, and water to bind the dough. This is used mainly in tarts. It is also the pastry that is used most often in making a quiche. The process of making pastry includes mixing of the fat and flour, adding water, and rolling out the paste. The fat is mixed with the flour first, generally by rubbing with fingers or a pastry blender, which inhibits gluten formation by coating the gluten strands in fat and results in a short (as in crumbly; hence the term shortcrust), tender pastry. A related type is the sweetened sweetcrust pastry, also known as *pâte sucrée,* in which sugar and egg yolks have been added (rather than water) to bind the pastry.

Flaky Pastry

Flaky pastry is a simple pastry that expands when cooked due to the number of layers. It bakes into a crisp, buttery pastry. The "puff" is obtained by the shard-like layers of fat, most often butter or shortening, creating layers which expand in the heat of the oven when baked.

Puff Pastry

Puff pastry has many layers that cause it to expand or "puff" when baked. Puff pastry is made using flour, butter, salt, and water. The pastry rises up due to the water and fats expanding as they turn into steam upon heating. Puff pastries come out of the oven light, flaky, and tender.

Choux Pastry

Choux pastry is a very light pastry that is often filled with cream. Unlike other types of pastry, choux is in fact closer to a dough before being cooked which gives it the ability to be piped into various shapes such as the eclair and profiterole. Its name originates from the French *choux*, meaning cabbage, owing to its rough cabbage-like shape after cooking.

Choux begins as a mixture of milk or water and butter which are heated together until the butter melts, to which flour is added to form a dough. Eggs are then beaten into the dough to further enrich it. This high percentage of water causes the pastry to expand into a light, hollow pastry. Initially, the water in the dough turns to steam in the oven and causes the pastry to rise; then the starch in the flour gelatinizes , thereby solidifying the pastry.[1] Once the choux dough has expanded, it is taken out of the oven; a hole is made in it to let the steam out. The pastry is then placed back in the oven to dry out and become crisp. The pastry is filled with various flavors of cream and is often topped with chocolate. Choux pastries can also be filled with ingredients such as cheese, tuna, or chicken to be used as appetizers.

Phyllo

Phyllo is a paper-thin pastry dough that is used in many layers. The phyllo is generally wrapped around a filling and brushed with butter before baking. These pastries are very delicate and flaky.

Hot Water Crust Pastry

Hot water crust pastry is used for savoury pies, such as pork pies, game pies and, more rarely, steak and kidney pies. Hot water crust is traditionally used for making hand-raised pies. The usual ingredients are hot water, lard and flour, the pastry is made by heating water, melting the fat in this, bringing to the boil, and finally mixing with the flour. This can be done by beating the flour into the mixture in the pan, or by kneading on a pastry board. Either way, the result is a hot and rather sticky paste that can be used for hand-raising: shaping by hand, sometimes using a dish or bowl as an inner mould. As the crust cools, its shape is largely retained, and it is filled and covered with a crust, ready for baking. Hand-raised hot water crust pastry does not produce a neat and uniform finish, as there will be sagging during the cooking of the filled pie, which is generally accepted as the mark of a hand-made pie

Priciples of Pastry

Pastry Terminology

- Blind Crust:
  - Baking a pie crust without filling.
  - Must prick the bottom of the pie with a fork to prevent blistering during baking or use dry beans or pastry beads to weigh down the pastry.
- Fluting: Decorative edge given to pastries.

Pastry Ingredients

1. Flour:

- Gives structure
- Too much: Tough pastry
- Too little: Gluten will not develop

2. Water:

- Provides moisture by producing steam
- Too much: Tough pastry
- Too little: Crumbly and difficult to roll

3. Salt:

- Flavour

4. Fat:

- Makes pastry tender
- Prevents gluten from overdeveloping
- Too much: Tough pastry
- Too little: Crumbly

Proper Fats For Pastry:

- Shortening: flaky crust

- Butter:
  a) Contributes a delicious flavor.
  b) Does not produce as flaky of a crust.
  c) Low melting point makes it difficult to work with

Can Do a Combination of the Two

Fats for Pastry

- Lard: Flakiest crust, however flavour is not suitable for fruit pies.
- Oil: Crust becomes extremely fragile, without flakiness.

Handling the Dough

- Too much handling =Tough pastry
  Causes gluten to overdevelop
- Handle gently at all times and as little as possible
- DO NOT:
- Overmix the dough when adding liquid
- Use the rolling pin vigorously
- Stretch the pastry into the pie plate

Characteristics of High-Quality Pastries

- TENDER
- FLAKY
- Layers of gluten will separate layers of fat and expand by steam
- Lightly and evenly browned
- Pleasing flavour

1 Derivatives: Frangipane, Creme Diplomate, Mousseline, Chiboust and Ligera

Like I mentioned before there are several derivative creams from the basic pastry cream.

- Frangipane = pastry cream 1 part + almond cream 2 parts
- Creme Diplomate = pasty cream + 3/4-point heavy cream
- Mousseline = pastry cream + cold butter, refrigerate, then + soft butter
- Chiboust = pastry cream + 50% Italian merengue + 1-2% gelatine
- Ligera = pastry cream + up to 60% fully beat heavy cream ¨a punto¨

Frangipane: Do a cream age of 4 ingredients – butter, powdered sugar, eggs and almond flour. (If you want to make the mixture economical because of the price of the almond flour, you can reduce the ratio to 4 to 1.)

Creme Diplomate: If you want to flavour the creme diplomat with chocolate, after you have boiled the mixture two minutes, add the cut-up chocolate and mix to emulsify it. Do not leave any clumps. Cover with plastic and put it in the refrigerator until it reaches 4C. Then take it out and the heavy cream beat to 3/4 point, then carefully fold the mixture until well mixed. Refrigerate.

Mousseline: If you want to flavour the mousseline, for example with a pistachio paste, add it at the beginning with the milk. Follow all the steps until boiling it for two minutes. Then move the mixture to a bowl and add the COLD butter, and then mix quickly and allow it to cool to 21C. At that point, put the mixture in a mixer's bowl and beat with SOFT butter.

Chiboust: Make the pastry cream move it to bowl, and add the activated gelatin. Refrigerate with plastic in contact with the cream. Take it out of the refrigerate, beat it again so it has a creamy texture, then do a pre-mix of Italian Merengue and carefully fold the mixture together.

Tips:

Each of these derivative creams are for a specific dessert. Try to apply them to the right one. For example, with the ***Gateau Saint Honore***, the Chiboust cream is used.

Always mix with carefully folding or you'll take out the air of the mixture

If you're flavouring with a herbs or spices or a pistachio paste, infuse the milk. If it is a softer flavor like chocolate, mix it after boiling the mixture.

Characteristics of Cakes

- Chief characteristics of cakes can be classified as **internal** and **external** characteristics.
- The external characteristics include volume, crust color, symmetry of form and character of crust.
- The internal characteristics include grain, crumb color and sensory parameters such as taste, aroma and texture.

Volume: It depends on consumer preference. It should be well risen with slight convex top surface and should not appear too small or too large for its weight.

Crust color: Pleasing golden brown color is desirable. Too dark or too light or dull color is not desirable. Crust must have a uniform color, free from dark streaks or sugar spots or grease spots.

Symmetry of form: Cakes should have symmetrical appearance. Peaking, crack on top surface, low sides, sunken or high center, burst, caved in bottom or uneven top are undesirable characteristics of cakes.

Character of crust: Crust should be thin and tender. It should not be rubbery, sticky or over moist, too tender, tough or busty crust indicates poor quality of cakes.

Internal Characteristics

- **Grain**: The grain is the structure formed by the extended gluten strands including the area they surround. Grain will vary according to type of cake. Uniformity of the size of cells and thin cell walls are desirable qualities. Coarseness, thick cell wall, uneven size of grains, large holes and tunnels are indicative of poor grain. Grains should not be too open or too close.

- **Color of Crumb**: It should be lively, lustrous and uniform color. It should be free from any streaks or dark patches. Grey, non uniform, dark, light or dull color crumb are undesirable.
- **Aroma**: Pleasant, rich, sweet and natural aroma is desired. Flat, misty, strong or sharp aroma is indicative of poor quality of cake.
- **Taste**: It should be pleasant, sweet and satisfying without any after taste or foreign taste. Salt and soda in excessive amounts affect the taste adversely.
- **Texture**: Texture denotes the pliability and smoothness of the crumb as felt by sense of touch. It depends on the physical condition of crumb and type of grain. A good texture is soft and velvety without weakness and should not be crumby. Rough, harsh, too compact, lumpy or too loose texture is not.

Balancing Cake Formula

- It is possible to change cake formulas, either to improve them or to reduce costs. However, ingredients and quantities can be changed only within certain limits.
- A cake formula whose ingredients fall within these limits is said to be "in balance." Knowing the limits, helps you not only modify recipes but also judge untested recipes and correct faults.

Ingredient Functions

- For the purpose of balancing cake formulas, we can classify cake ingredients according to four functions: toughness (or stabilizers), tenderizers, dryers, and moisteners (or moisturizers).
- The idea of formula balancing is that toughness should balance tenderizers and dryers should balance moisteners. If, for example, we increase the amount of toughness in a formula, we must compensate by also increasing the number of tenderizers.
- Many ingredients fill more than one function, sometimes even opposite functions. Egg yolks contain protein, which is a toughened, but they also contain fat, which is a tenderizer. The major cake ingredients act as follows:

Tougheners

- Provide structure: flour, eggs (whites and yolks).
- Tenderizers provide softness or shortening of protein fibres: sugar, fats (including butter, shortening, and cocoa butter), and chemical leaveners.

Moisteners

- Provide moisture or water: water, liquid milk, syrups and liquid sugars, eggs.

Dryers

- Absorb moisture: flours and starches, cocoa, milk solids.

Balancing Fat

Type or Shortened Cakes

- The general rules for balancing creaming-method cakes made with butter or regular shortening are as follows (all ingredient quantities are by weight):
- The sugar (a tenderizer) is balanced against the flour (a toughener). In most creaming method cakes, the weight of sugar is less than or equal to the weight of flour.
- The fat (a tenderizer) is balanced against the eggs (tougheners).
- The eggs and liquids (moisteners) are balanced against the flour (a dryer).
- Balancing one ingredient against another, as indicated in the preceding guidelines, means that if one ingredient is increased or decreased, then the balancing ingredients must also be adjusted. For example, if the fat is increased, then the eggs must be increased to keep the formula in balance.
- A common practice in balancing a formula is to decide on the sugar/flour ratio and then balance the rest of the ingredients against these. The following guidelines are helpful in this regard:
  a) If eggs are increased, increase the shortening.
  b) If extra milk solids are added as enrichment, add an equal weight of water.
  c) If cocoa is added, add water equal in weight to 75 to 100% of the cocoa.
  d) If cocoa or bitter chocolate is added, increase the amount of sugar to as much as 180% of the weight of the flour in high-ratio cakes, and to over 100% of the weight of the flour in creaming-method cakes. This is to account for the starch content of the cocoa and chocolate.
  e) In cakes to be baked in very large units, use less liquid because less water will evaporate during baking.
  f) If a liquid sugar is added (honey, corn syrup, etc.), reduce other liquids slightly.
  g) If large quantities of moist ingredients, such as applesauce or mashed bananas, are added, reduce the liquid. Extra-large additions of moist ingredients may also require increasing the flour and eggs.
  h) Use less baking powder with creamed batters than two-stage batters because the creamed batters get more aeration in the creaming stage.

Common Cake Faults and their Causes

Fault	Causes
Volume and Shape	
Poor volume	Too little flour
	Too much liquid
	Too little leavening

Fault	Causes
	Oven too hot
Uneven shape	Improper mixing
	Batter spread unevenly
	Uneven oven heat
	Oven racks not level
	Cake pans warped
CRUST	
Too dark	Too much sugar
	Oven too hot
Too light	Too little sugar
	Oven not hot enough
Burst or cracked	Too much flour or flour too strong
	Too little liquid
	Improper mixing
	Oven too hot
Soggy	Under baked
	Cooling in pans or with not enough ventilation
	Wrapping before cool
Texture	
Dense or heavy	Too little leavening
	Too much liquid
	Too much sugar
	Too much shortening
	Oven not hot enough
Coarse or irregular	Too much leavening
	Too little egg
	Improper mixing
Crumbly	Too much leavening
	Too much shortening
	Too much sugar
	Wrong kind of flour
	Improper mixing
Tough	Flour too strong
	Too much flour
	Too little sugar or shortening

Fault	Causes
Poor Flavor	Over mixing Poor-quality ingredients Poor storage or sanitation Unbalanced formula

3. Chocolate

Smooth in texture, rich in taste, chocolate is loved by almost everyone; even the gods, as its scientific name "Theobroma cacao" confirms. Theobroma means the food of the gods in Greek. Today much of the worlds cocoa crops come from Africa and Brazil

Cocoa Pods

The Cacao tree fruits grow direct on the trunk and on larger branches of the tree. This pattern is known as "cauliflory". The early illustrations of the Cacao tree from Europe often showed the tree with fruits at the end of smaller branches - just like many other fruit trees have it.

The Cacao tree needs to mature for 10 years, before the best fruit can be harvested and reaches 30-45 feet in height. However, the tree is usually cut to 18-20 feet to make harvesting process easier.

The colour of the ripe Cacao fruits varies greatly between yellowish brown to almost red. It takes 4-8 months for a tree to produce ripe fruit depending on the variety of the Cacao. Each fruit contains about 30 oval shaped seeds (the cacao beans) surrounded by a sweet-sour fruity pulp-like substance.

On average each Cacao bean consists of 54% cacao butter, 11.5% protein, 9% starch and hundreds of other elements including Theobromine, caffeine and aromatic oils.

The water accounts to about 5% of the Cacao bean. Theobromine has an effect on human organism very similar to that of caffeine. This is why Chocolate should be consumed in moderation, especially by children.

Cocoa Processing

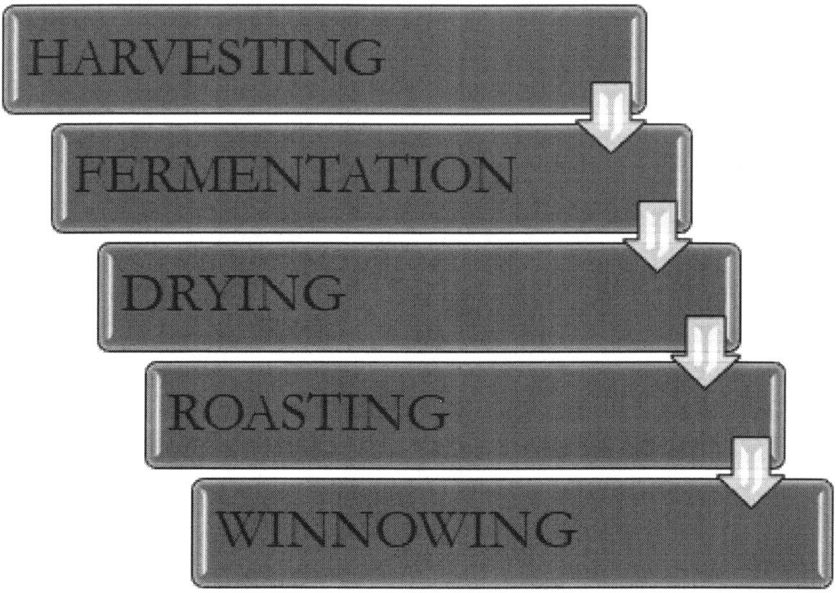

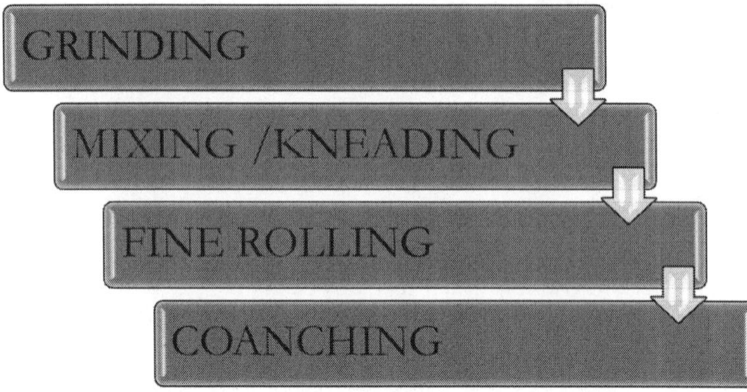

Harvesting

The Cacao is harvested throughout the year, but main seasons are November to January and May to July.

The harvesting is done by hand cutting the Cacao fruit down with sharp Machetes. It impossible to use machines for harvesting, because of the various sizes of the fruit and because the tree can not be damaged, since it is continuously producing new flowers and fruit. After the fruit has been cut down, it is opened and the seeds (beans) with its surrounding pulp are extracted.

Fermentation

- After the beans and pulp has been extracted, the fermentation takes place.
- The beans with pulp are placed in fermentation boxes or between banana leaves.
- They will stay there for 5 - 6 days during which the sugar from the beans will be turned into alcohol.
- The pulp will become a liquid and will drain out. During the fermentation the seeds will begin to germinate, soon to be killed by the high temperatures and increased acidity due to the fermentation process.
- It is during the fermentation, that the Cacao bean develops the brown colour of typical Cacao.

Drying

- After the fermentation, the beans will have to dry.
- The water content is about 60% and it will have to be lowered to 7% before the beans can be traded and stored.
- In most places the drying is done by the sun. The beans are spread out in 2 inch thick layers and are dried for about 14 days.
- Industrial producers use hot air or heat from burning wood to dry the beans.
- This is the fast way to dry them, but they will get undesirable, somewhat smoky flavour that will not disappear during the rest of the process.
- The fermented and dried cocoa beans as they are supplied to factories.

Roasting

- The next step in the preparation of Chocolate is the roasting.
- The roasting which involves temperatures of 210F, lasts for anything between 10 and 115 minutes.
- The roasting is necessary for the development of the unique Cacao flavour and aroma.
- It is a very critical process that must be closely monitored at all times.
- If the beans are roasted at low temperature and little time, the fruitiness of the Cacao bean will be well preserved.
- If the beans are of the poor quality they will need to be roasted longer and at higher temperatures.
- Roasting longer will increase the bitterness of the beans making them less aromatic or perhaps, if the beans are over-roasted, they become completely useless.

Winnowing

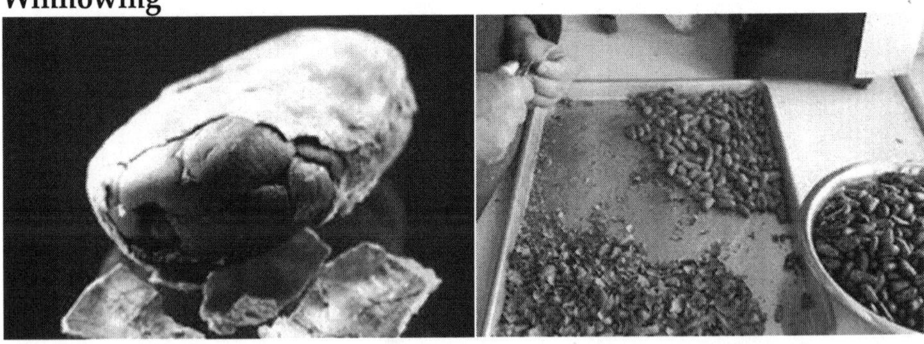

- The roasting is followed by the winnowing where the Cacao bean shells are removed.

Grinding

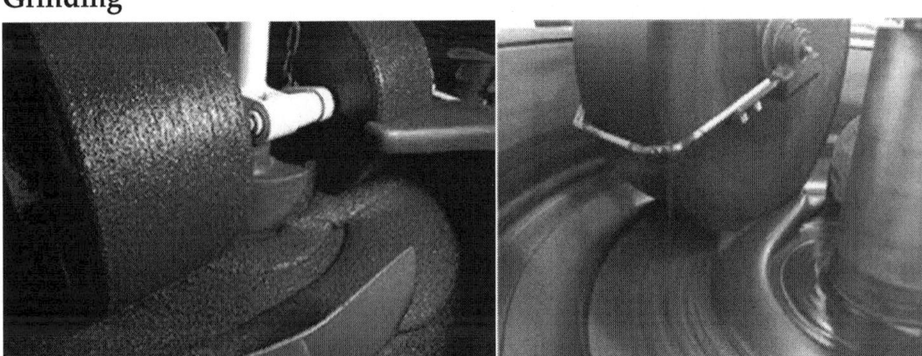

- Now the beans are ready to be grinded. There are two stages of the grinding process.
- In the first stage, the beans are grinded between large stones. This stage will produce a liquid mass called cacao liquor.

- From here, there are two possible ways to continue: either the cacao liquor is further processed into Cacao butter and cacao powder (usually only done with low-quality beans) or the liquor is processed into Chocolate.
- To produce Chocolate (the final product most of us crave for so much and so often!) sugar is added. The size of the particles in the chocolate mass is now about 100 microns particles .
- To avoid "grainy" taste in the finished product, another grinding takes place. This time the particle size is about 18 microns: virtually Cocoa dust!

Conching

- After the grinding, the process called conching takes place. During the conching the Chocolate liquid mass is stirred and mixed at a temperature of about 180F.
- During this process the friction between the sugar- and the cacao-particles occurs. This causes further "polishing" of the Cacao particles thus contributing to the smooth taste of the finished chocolate.

Different Forms of Cocoa and Chocolate Available

Forms of Cocoa

Natural Cocoa

- Dry powder that remains after part of the cocoa butter has been removed from chocolate liquor.
- Natural cocoa is somewhat acidic.
- When it is used in such products as cakes, it is possible to use baking soda (which reacts with acid) as part of the leavening power.

Dutched Cocoa

- Dutch process cocoa, or dutched cocoa, is processed with an alkali.
- It is slightly darker, smoother in flavour, and more easily dissolved in liquids than natural cocoa.
- It is generally neutral or even slightly alkaline.
- It does not react with baking soda. Instead, baking powder is used as the sole leavening agent.

BAKING SODA NEEDED TO BALANCE THE ACIDITY OF TYPICAL COCOA PRODUCTS

	AMOUNT OF BAKING SODA PER LB	AMOUNT OF BAKING SODA PER KG
Natural cocoa	1.25 oz	80 g
Dutched cocoa	0	0
Unsweetened chocolate	0.8 oz	50 g
Sweet chocolate	0.4 oz	25 g

- If you are substituting dutched for natural cocoa, you must increase the baking powder by 1 ounce (30 g) for each 1/2 ounce (15 g) soda omitted.
- If not, enough soda is used in chocolate products, the color of the finished product may range from light tan to dark brown, depending on the quantity used.
- If too much is used, the colour will be reddish brown.
- When switching from one kind of cocoa to another, you may have to adjust the soda in your recipes.

Cocoa Butter

Cocoa butter is the fat pressed out of chocolate liquor when cocoa is processed.

Its main use in the bakeshop is to thin melted couverture to a proper consistency.

Forms of Chocolate Available

Unsweetened or Bitter Chocolate

- Unsweetened chocolate is straight chocolate liquor.
- It contains no sugar and has a strongly bitter taste so known as Bitter Chocolate
- As it is moulded in blocks, it is also referred to as block cocoa or cocoa block.
- It is used to flavor items that have other sources of sweetness.
- In some less expensive brands of unsweetened chocolate, some of the cocoa butter may be replaced by another fat.

Sweetened Dark Chocolate

- Sweetened dark chocolate is bitter chocolate with the addition of sugar and cocoa butter in varying proportions.
- Based on the amount of sugar added, dark chocolate is divided into the following categories:
- **Sweet chocolate** has the **highest** percentage of **sugar** and the **lowest** percentage of **chocolate liquor**.

- A product labelled sweet chocolate may contain as little as **15%** chocolate liquor, although it may contain as much as **50%**. Because of the low cocoa content, this product is used primarily in **inexpensive** candies, not in the finest chocolate work.
- **Semisweet chocolate** must contain a minimum of **35%** chocolate liquor, but it usually contains about **50 to 65%**. It is widely used in confectionary, pastry, and dessert products.
- **Bittersweet chocolate** contains the **highest** proportion of **chocolate liquor**, usually from **65 to 85%**. Products at the upper end of this range are sometimes labelled **extra bittersweet**. Because of its **higher cocoa** content, bittersweet chocolate is used in the **finest** baked goods, pastries, desserts, and confections.
- **Note**: According to government regulations, there is **no difference** between semisweet and bittersweet chocolate. Both have a minimum chocolate liquor content of **35%**. However, most manufacturers use the term **bittersweet** for their products with **higher cocoa** content.

Points to consider while using different types of Chocolate

- When a sweetened dark chocolate is required in a formula, semisweet chocolate is usually specified.
- **Bittersweet chocolate** is specified when a **good grade** of chocolate with a high chocolate liquor content is essential for the **best results**.
- **Sweetened** chocolate has, on average, **only half** the chocolate content of **bitter** (unsweetened) chocolate, it is usually **not economical** to add to products that are already **highly sweetened**, because twice as much will be needed. For example, it is better to use bitter chocolate when making chocolate fondant from plain white fondant.
- Good-quality chocolate products—including not only dark chocolate but also milk chocolate and white chocolate—are often called **couverture**, which means "**coating**" in **French**.
- When couverture is used to coat candies, cookies, and other products, the chocolate must be prepared by a process called **tempering**.
- Tempering involves **carefully melting the chocolate** without letting it get too warm, then bringing the temperature back down to a certain level.

Coating Chocolate

- **Less expensive** chocolates, which have **part of** the **cocoa butter** replaced by **other fats**, are easier to handle and do not require tempering.
- They do not have the flavor and eating qualities of good chocolate.
- These products are sold under such names as cookie coating, cake coating, baking chocolate, coating chocolate, and compound chocolate.
- Do not confuse coating chocolate with couverture. These two products are entirely different, even though couverture means "coating." It would be less confusing if this lower-quality chocolate were referred to only as baking chocolate, without using the word coating.

Milk Chocolate

- Milk chocolate is sweet chocolate to which milk solids have been added.
- It is usually used as a coating chocolate and in various confections.
- Milk chocolate is seldom melted and then incorporated in batters because it contains a relatively low proportion of chocolate liquor.

White Chocolate

- White chocolate consists of cocoa butter, sugar, and milk solids.
- It is used primarily in confectionery.
- Technically, it should not be called chocolate, because it contains no cocoa solids.
- Some inexpensive brands, in which another fat is substituted for the cocoa butter, not only contain no cocoa solids, but none of the other components of chocolate either. These products do not deserve the name chocolate at all.

Chocolate Tempering

Procedure: Tabling Method

Caution: In all stages of this procedure, do not let even a trace of moisture come in contact with the chocolate.

1. With a heavy knife, chop the chocolate into small pieces. Place the pieces in a dry stainless-steel bowl.
2. Set the bowl in a pan of warm water. Stir the chocolate constantly so it melts uniformly.
3. Continue stirring until the chocolate is completely melted and reaches the proper temperature, as indicated in the Critical Temperatures for Tempering Chocolate table.
4. Remove the bowl from the water bath. Wipe all traces of moisture from the bottom of the bowl, to avoid contaminating the chocolate.
5. Pour about two-thirds of chocolate onto a marble slab.
6. With a metal scraper and a spatula, spread the chocolate and quickly scrape it back together, continuing to mix the chocolate so it cools uniformly.
7. When the chocolate cools to the proper temperature (26°–29°C), it will become thick and pasty. Quickly scrape it back into bowl with remaining melted chocolate.
8. Mix and reheat the chocolate over hot water to the proper rewarming temperature (29°–32°C, depending on chocolate). Do not warm it above the recommended temperature.
9. The chocolate is now ready for use.

1

2

3

4

Procedure Variation

- Instead of working the chocolate on a marble slab as indicated, some chefs prefer to set the entire bowl over cold water and stir until chocolate cools to proper temperature.
- Then they rewarm the chocolate. This method may not produce chocolate with as good a temper, and it raises the risk of getting water in the chocolate.
- However, it has the advantage of speed.

Procedure: Seeding Method

1. Chop the chocolate to be melted into small pieces, as in method 1.
2. Cut fine shreds or shavings from a block of tempered chocolate and set them aside.
3. Melt the chopped chocolate as in method 1.
4. Remove the melted chocolate from the water bath. Stir in some of the shaved chocolate.
5. When these shavings are nearly all melted, add a few more shavings. Continue adding and stirring until the melted chocolate is cooled to the proper tempering point. Do not add shavings too fast, or they may not all melt.
6. Rewarm the chocolate as in method 1.

Chocolate Designs

Chocolate Cutouts: Procedure

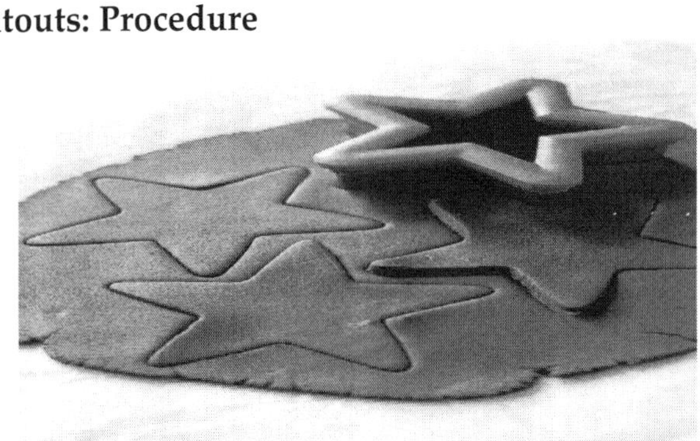

1. Mix dark and white couverture very lightly to marble.
2. Polish a sheet of acetate with cheesecloth or cotton wool. Pour some of the chocolate onto a sheet of acetate.
3. Lift the acetate and tilt it from side to side so the chocolate coats the entire sheet. Or spread into a thin layer with a palette knife.
4. When the chocolate clouds over and is set but not hard, cut out desired shapes with a small, sharp knife or cutter, or use a metal cutter, slightly warmed. Do not attempt to remove the cut-outs at this point.

5. Let stand until the chocolate is hard and the acetate peels away easily from the chocolate.
6. Broken pieces of dark chocolate streaked with white and cut-outs of marbled white and milk chocolate.

Some of the easiest and most popular techniques:

- **Spread with one colour of chocolate**, and then scrape with a comb scraper, using the same technique as for ribbon sponge. Then cover with another colour.
- **To marble chocolate**, add a little tempered white couverture to tempered dark couverture and mix just until the white shows streaks against the dark. Then coat the acetate.
- **Use chocolate transfer sheets**, which are commercially made acetate sheets with designs on them. Use as you would a plain acetate sheet, making sure the raised or textured side—the side that has the design on it—is facing up. After the chocolate has set and the acetate is peeled off, the design remains on the chocolate. Custom-made transfer sheets, embossed with a company's name or logo.
- With a paper cone

Curled Cutouts: Procedure

1. Place a transfer sheet, rough side up, on a work surface, preferably a marble slab.
2. Spread with a layer of tempered chocolate
3. When the chocolate is set but not hard, cut into triangles or other desired shapes
4. Lay a sheet of parchment over the chocolate and roll up diagonally
5. Set the roll in a curved mould to secure it while it hardens
6. When the chocolate has hardened, peel off the acetate and carefully separate the cut-outs

Chocolate Strips Wood Grain Pattern: Procedure

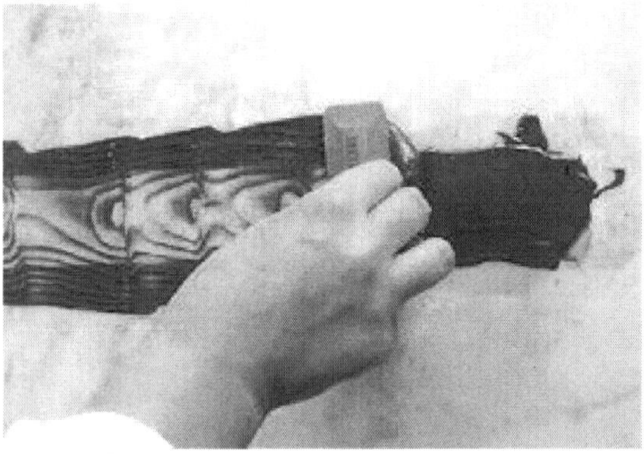

1. Drizzle a little tempered dark chocolate onto an acetate strip and, with a palette knife, spread it into a thin layer covering the strip
2. Scrape wood-grain tool down the length of strip, rocking it back and forth to make the grain pattern let stand a few minutes to allow the chocolate to set.
3. Spread the strip with a layer of tempered White couverture
4. Lift the acetate strip and run your fingers along both edges to remove excess chocolate.
5. The acetate side of the strip shows the pattern.
6. Fit the strip into a ring mould and allow it to set

Chocolate Bows: Procedure

Chocolate strips made into teardrops (like bows, but larger) are used for plating presentations of desserts such as chocolate mousse.

1. Cut strips of acetate to the desired size.
2. Spread with tempered chocolate, either in a single color or in a pattern of two colors.
3. Allow the chocolate to cloud and partially set, then bend into a teardrop shape and fasten the ends together with a paper clip.

4. If necessary, pipe a little tempered chocolate with a paper cone into the joint between the ends of the chocolate to strengthen the seam. Let set until hardened.
5. Peel the acetate from the bows after they are completely set
6. Cut the ends to points so they will fit together on the box
7. To hold the bows in place, pipe a little tempered chocolate with a paper cone
8. Fit the bow in place. Hold steady until the chocolate sets

Curled Strips: Procedure

1. Spread a layer of chocolate onto an acetate strip
2. Scrape along the length of the strip with an icing comb
3. Using a paper cone filled with tempered chocolate, connect the lines of chocolate with small dots as shown
4. Carefully lift the acetate strip and twist it into a spiral. Set it into a curved mould to hold it securely while the chocolate hardens.
5. When chocolate is hard, peel off the acetate
6. Trim off any excess to make the finished curl

Chocolate Cigarettes

1. Melted chocolate can be used for these decorations.
2. Tempered chocolate may also be used but is not necessary.
3. Spread the chocolate in a long strip on a slab of marble. Allow the chocolate to set. It should be completely set but not hard and brittle.

4. If it becomes too hard, warm it slightly by rubbing it with the palm of your hand.
5. Hold a metal scraper and push it forward so the chocolate curls up ahead of the metal edge.

Chocolate Curls and Shavings

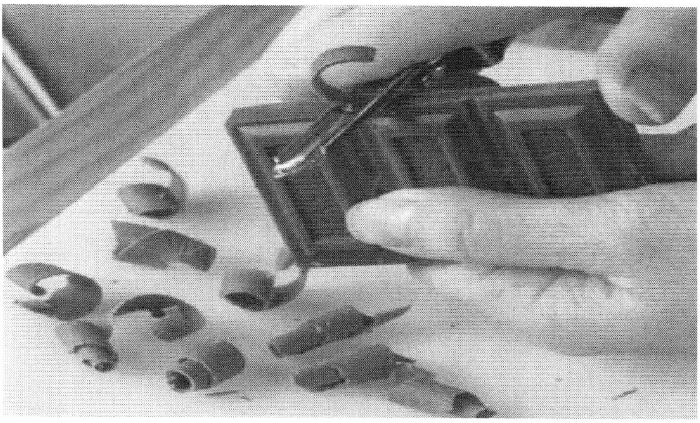

1. Score the strip of chocolate with the point of a knife
2. Then scrape up the curls with the knife

Soft Chocolate Fans: Procedure

Initial procedure for Chocolate Fans and Strips.

- Heat a clean half-sheet pan at 325°F (160°C) for 4 minutes.
- Use a clean, unwrapped pan, preferably one reserved for chocolate work only.
- The tray should be warm but not too hot to handle, to spread thinner coat of chocolate
- Turn the sheet upside down and spread melted chocolate in a thin, even layer on the bottom of the pan.
- Let stand until the chocolate looks cloudy, then refrigerate until set.
- Remove from the refrigerator and let warm to room temperature.

1. Scrape the chocolate as for strips, except hold your thumb against one corner of the scraper so the strip bunches or gathers on one side
2. Carefully neaten the ruffles or folds of the fan as necessary

Soft Chocolate Strips: Procedure

1. Push the scraper against the tray and, with the other hand; lift off the strip of chocolate as the scraper releases it
2. Handle the strip of chocolate very lightly
3. Coat the sides of the cake
4. Make more strips for the top
5. Arrange the strips on the top
6. Finish the cake with coiled strips
7. Touch them as little as possible to prevent melting

Chocolate Petals: Procedure

- Like soft strips and fans, chocolate petals are made by scraping chocolate off a flat surface.

The scraping tool is a round cutter.

1. Spread a thin layer of chocolate on a marble slab or other flat surface

2. When the chocolate is set but not yet hard and brittle, hold the bottom edge of a round cutter as shown and scrape up curled petals by pulling the cutter toward you.
3. The resulting cup-shaped petals can be used for a variety of decorative purposes

Piping Chocolate

1. With the use of a paper cone, tempered chocolate can be piped directly onto the dessert or onto parchment paper in small designs and left to harden.
2. After they are set, the decorations can be removed from the paper and placed on the desserts.
3. Decorations can be made during slack hours and stored until needed.
4. Ordinary tempered chocolate is adequate for small, fine decorations, such as those for petits fours, but it is too thin for most other piped decorations.
5. To make piping chocolate, add a very little warm simple syrup to the tempered chocolate. This will thicken the chocolate immediately. Stirring constantly, add more syrup, very slowly, until the chocolate is thinned to piping consistency.

4. SUGAR

Sugar Cooking Stages

Testing the temperature with a candy thermometer is the most accurate way to determine the doneness of a syrup. In the old days, a syrup was tested by dropping a little of it into a bowl of cold water and checking the hardness of the cooled sugar.

The stages of doneness were given names that described their hardness. The Stages of Doneness in Sugar Cooking table here lists these stages of sugar cooking. Please note that the names for the stages are not absolute; different sources may use slightly different names. In fact, all such listings are misleading because they suggest the syrup jumps from one stage to the next. Actually, of course, it changes gradually as the water is boiled off. For this reason, it is best to rely on the thermometer and not worry too much about the names.

Stages of Doneness In Sugar Cooking

	Temperature	
Stage	°F	°C
Thread	230	110
Soft ball	240	115
Firm ball	245	118
Hard ball	250-260	122-127
Small crack	265-270	130-132
Crack	275-280	135-138
Hard crack	290-310	143-155
Caramel	320-340	160-170

Caramelised Sugar Preparation

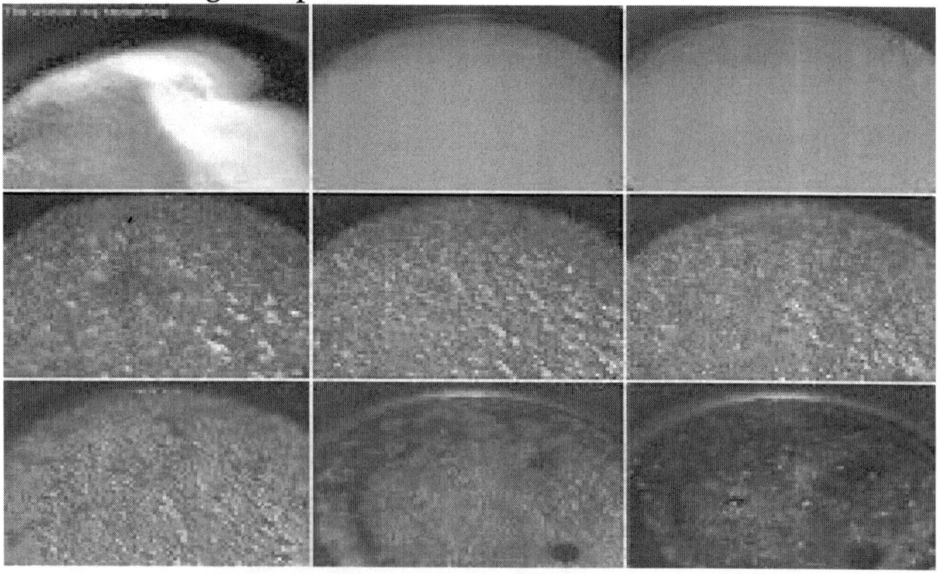

Syrup Strength

Syrup strength is an indication of the concentration of sugar in a solution. In small concentrations, sugar can be dissolved in water simply by stirring it in. With larger concentrations, however, we boil the syrup, because at higher temperatures, the sugar dissolves more quickly. Also, boiling water can hold more sugar in solution than cold water can.

Once the sugar has dissolved, we can increase the concentration of sugar by continuing to boil the syrup so water gradually evaporates. As the water is boiled off, the temperature of the syrup gradually rises. When all the water has evaporated, we are left with melted sugar. The sugar then begins to caramelize, or turn brown and change flavor. If heating continues, the sugar continues to darken and then burn.

A syrup cooked to a high temperature is harder when it is cooled than is a syrup cooked to a lower temperature. For example, a syrup cooked to 240°F (1 1 5 ° C) forms a soft ball when cooled . A syrup cooked to 300°F (150°C) is hard and brittle when cooled.

One pint (1 pound) water is enough to dissolve 3 or 4 pounds of sugar. There is no point in adding more water than you need for a particular purpose because you will just have to boil it off again. Pure, clean granulated sugar is used to make syrups. Impurities cloud the syrup and form a scum or foam on the syrup as it is being boiled. Any scum should be carefully skimmed off.

Sugar for Desert and Presentation

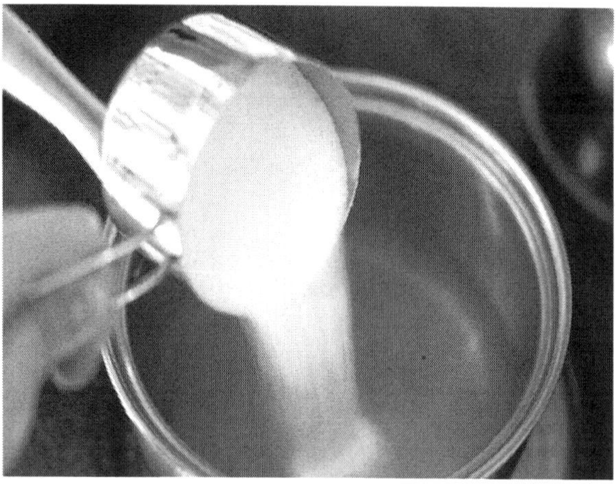

Syrups for the Bakeshop

Two basic syrups are kept in stock in the bakeshop and used in a variety of ways. **Simple syrup,** also known as **stock syrup,** is a solution of equal weights of sugar and water. It is used for such purposes as diluting Fondant and for preparing a variety of dessert syrups.

Dessert syrup, also called cake syrup, is simply a flavored simple syrup. It is used to moisten and flavor sponge cakes and various desserts, such as Babas au Rum.

The concentration of both these syrups may be varied to taste. Some chefs prefer a sweeter syrup for some purposes, such as 1 part water to 1 1/2 parts sugar. Others use a less sweet syrup, such as 2 parts water to 1 part sugar.

5. Icing and Topping

Icings, also called frostings, are sweet coatings for cakes and other baked goods. Icings have three main functions:

- They contribute flavor and richness.
- They improve appearance.
- They improve keeping qualities by forming protective coatings around cakes.

They are eight basic types of icings and other cake coatings:

1. Fondant
2. Buttercreams
3. Foam-type icings
4. Fudge-type icings
5. Flat-type icings
6. Royal or decorator's icing
7. Glazes
8. Rolled coatings

Use top-quality flavorings for icings so they enhance the cake rather than detract from it. Use moderation when adding flavors and colors. Flavors should be light and delicate. Colors should be delicate pastel shades—except chocolate, of course.

Fondant

Fondant is a sugar syrup that is crystallized to a smooth, creamy white mass. It is familiar as the icing for napoleons, éclairs, petits fours, and some cakes. When applied, it sets up into a shiny, nonstick coating.

Because it is difficult to make in the bakeshop, fondant is almost always purchased already prepared, either in the ready-to-use moist form or the dry form, which requires only the addition of water. In an emergency (for instance, if you run out of fondant and there is no time to get more from your supplier), flat icing can be substituted, although it will not perform as well. For those who wish to try making fondant, a formula is included here. The purpose of the glucose or cream of tartar is to invert some of the sugar in order to get the right amount of crystallization. If none is used, the syrup will set up to be too unworkable, and it will not be smooth and white. When an excess of glucose or cream of tartar is added, not enough crystallization will take place and the fondant will be too soft and syrupy. Also, if the hot syrup is disturbed before it cools sufficiently (step 6 in the procedure), large crystals will form and the fondant will not be smooth and shiny.

Another type of fondant, rolled fondant, is handled and applied differently from poured fondant.

Procedure and Guidelines: Using Fondant

1. Heat the fondant over a warm-water bath, stirring constantly, to thin it and make it pourable. Do not heat it over 100°F (38°C), or it will lose its shine.
2. If it is still too thick, thin it with a little simple sugar syrup or water.
3. Add flavorings and colorings as desired.
4. To make chocolate fondant, stir melted unsweetened chocolate into the warm fondant until the desired color and flavor are reached (up to about 3 oz. bitter chocolate per lb. of fondant, or 190 g per kg). Chocolate will thicken the fondant, so the icing may require more thinning with sugar syrup.
5. Apply the warm fondant by pouring it over the item or by dipping items into it.

Buttercreams

Buttercream icings are light, smooth mixtures of fat and sugar. They may also contain eggs to increase their smoothness or lightness. These popular icings for many kinds of cakes are easily flavored and colored to suit a variety of purposes.

There are many variations of buttercream formulas. We cover five basic kinds in this chapter:

1. **Simple Buttercreams** are made by creaming together fat and confectioners' sugar to the desired consistency and lightness. A small quantity of egg whites, yolks, or whole eggs may be whipped in. (For safety, use only pasteurized eggs.) Some formulas also include nonfat milk solids.

   Decorator's buttercream (sometimes called rose paste) is a special type of simple buttercream used for making flowers and other cake decorations. It is creamed only a little, at low speed, as too much air beaten into it would make it unable to hold delicate shapes. Because shortening has a higher melting point than butter, it is often used as the only fat in decorator's buttercream, to give maximum stability to the finished décor. However, when possible, a little butter may be included to improve the flavor.

2. **Meringue-Type Buttercreams** are a mixture of butter and meringue. These are very light icings. The most frequently made of these kinds of buttercreams is Italian buttercream, made with Italian Meringue. Swiss Meringue can also be used as the base for buttercream.

3. **French Buttercreams** are prepared by beating a boiling syrup into beaten egg yolks, and whipping to a light foam. Soft butter is then whipped in. These are very rich, but light, icings.

4. **Pastry Cream-Type Buttercream**, in its simplest form, is made by mixing together equal parts thick pastry cream and softened butter, and whipping until light. If more sweetness is desired, sifted confectioners' sugar may be mixed in. To give it the necessary body, a little gelatin is added. This type of preparation is better suited for use as a cake filling rather than an exterior icing.

5. **Fondant-Type Buttercream** is simple to make with only a few ingredients on hand. Simply cream together equal parts fondant and butter. Flavor as desired. Butter, especially unsalted butter, is the preferred fat for buttercreams because of its flavor and melt-in-the-mouth quality. Icings made with shortening only can be unpleasant because the fat congeals and coats the inside of the mouth, where it does not melt. However, butter makes a less stable icing because it melts so easily. There are two ways around this problem:

   - Use buttercreams only in cool weather.
   - Blend a small quantity of emulsifier shortening with the butter to stabilize it.

Buttercreams may be stored, covered, in the refrigerator for several days. However, they should always be used at room temperature in order to have the right consistency. Before using, remove buttercream from the refrigerator at least

1 hour ahead of time and let it come to room temperature. If it must be warmed quickly, or if it curdles, warm it gently over warm water and beat it well until smooth.

Flavoring Buttercreams

Buttercreams may be combined with many flavorings, making them versatile and adaptable to many kinds of cakes and desserts. The quantities given in the following variations are suggested amounts for each 1 pound (500 g) buttercream. In practice, flavorings may be increased or decreased to taste, but avoid flavoring icings too strongly. Unless the instructions say otherwise, simply blend the flavoring into the buttercream.

1. **Chocolate.** Use 3 ounces (90 g) semisweet dark chocolate. Melt chocolate and cool slightly. (Chocolate must not be too cool or it will solidify before completely blending with the buttercream.) Blend with about one-quarter of the buttercream, then blend this mixture into the rest. If your buttercream base is very sweet, use 1 1/2 ounces (45 g) unsweetened chocolate instead of the sweet chocolate.
2. **Coffee.** Use 2/3 fluid ounces (20 mL) coffee compound (coffee flavoring), or 1 1/2 tablespoon (5 g) instant coffee dissolved in 1/2 ounce (15 mL) water.
3. **Marron (Chestnut).** Use 8 ounces (250 g) chestnut purée. Blend with a little of the buttercream until soft and smooth, then blend this mixture into the remaining buttercream. Flavor with a little rum or brandy, if desired.
4. **Praline.** Use 2 to 3 ounces (60 to 90 g) praline paste. Blend with a little of the buttercream until soft and smooth, then blend this mixture into the remaining buttercream.
5. **Almond.** Use 6 ounces (180 g) almond paste. Soften almond paste with a few drops of water. Blend in a little of the buttercream until soft and smooth, then blend this mixture into the remaining buttercream.
6. **Extracts And Emulsions** (orange, lemon, etc.). Add according to taste.
7. **Spirits And Liqueurs.** Add according to taste. For example: kirsch, orange liqueur, rum, brandy.

Foam-Type Icings

Foam icings, sometimes called boiled icings, are simply meringues made with a boiling syrup. Some also contain stabilizing ingredients like gelatin. Foam icings should be applied thickly to cakes and left in peaks and swirls. These icings are not stable. Therefore, regular boiled icing should be used the day it is prepared. Marshmallow icing should be made just before using and applied while still warm, before it sets.

Plain Boiled Icing

Follow the recipe for Italian Meringue, but include 2 ounces (60 g) corn syrup with the sugar and water for the boiled syrup. Flavor the icing to taste with vanilla.

Marshmallow Icing

Soak 8 g gelatine in 45 mL cold water. Warm the water to dissolve the gelatine. Prepare plain boiled icing. Add the dissolved gelatine to the icing after adding the hot syrup. Scrape down the sides of the bowl to make sure that the gelatine is evenly mixed in. Use while still warm.

Chocolate Foam Icing and Filling

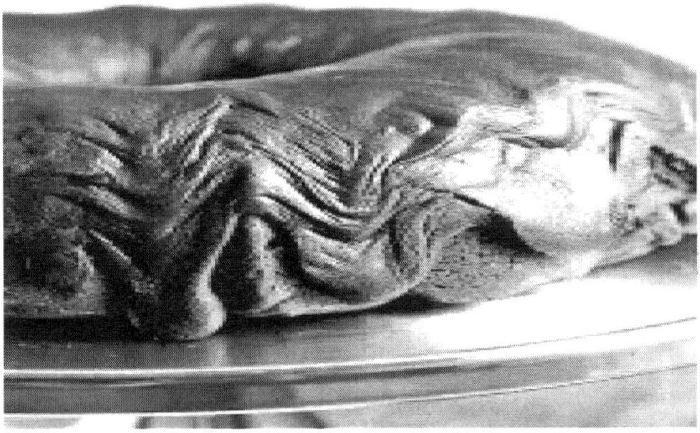

Prepare boiled icing. After the syrup has been added, blend in 5 ounces (150 g) melted, unsweetened chocolate.

FU Dge-Type Icings

Fudge-type icings are rich and heavy. Many of them are made somewhat like candy. Their predominant ingredient is sugar, and they contain less fat than buttercreams. Fudge icings may be flavored with a variety of ingredients, and are used on cupcakes, layer cakes, loaf cakes, and sheet cakes.

Fudge icings are stable and hold up well on cakes and in storage. Stored icings must, however, be covered tightly to prevent drying and crusting.

To use stored fudge icing, warm it in a double boiler until it is soft enough to spread.

Royal Icing

Royal icing, also called decorating or decorator's icing, is similar to flat icings except it is much thicker and made with egg whites, which make it hard and brittle when dry. It is used almost exclusively for decorative work. Pure white royal icing is most often used, but it may also be colored as desired. Because it consists mostly of confectioners' sugar, it is sweet but has little taste.

Royal icing dries easily and quickly, which makes it useful for fine decorations; but it also requires special handling and storage. Cover it tightly whenever it is not in use. For even greater protection from drying, place a clean, damp towel on the surface of the icing, and then cover the container tightly with plastic film. If any icing dries or crusts on the sides of the container during storage, remove the dried sections carefully, so that it does not fall back into the moist icing. Dried particles can clog the tips of paper cones and writing tubes.

Using a paper cone or, even better, a pastry bag fitted with a writing tip (a pastry tip with a small, round opening), royal icing can be piped into designs onto parchment or plastic and allowed to dry. They can then be lifted off carefully and stored in airtight containers for later use.

A second use for royal icing is called string work, in which delicate strands or filaments of icing are suspended between two attachment points, as illustrated in the decoration of the pastillage showpiece on page 663. This technique is also used on some styles of wedding cakes. To produce string work, touch the tip of the paper cone to the first attachment point, then pull the bag away while squeezing the cone with constant pressure. Allow the loop to drop to the desired length, then touch the tip to the second attachment point.

A third use of royal icing is flooding outlined areas of a design with colored icing. This technique requires much thinner icing than that used for string work. Thin the icing with water until a teaspoonful dropped into the bowl of icing flattens to a smooth surface in about 10 seconds. The first step in flooding is to draw an outline using medium-stiff royal icing, either white or colored, as desired. Pipe the outline on the desired surface, such as a sheet of acetate. Allow the outline to dry at least until the surface of the icing is firm. Using a pastry bag with a small plain tip, pipe thinned icing, colored as desired, in a line next to but not touching the inside of the dried outline. The icing should be thin enough to flow up to the outline. Continue to pipe along the inside edge of the icing until the area is filled with a smooth layer of icing. When dry, the design can be lifted off and placed on the surface of a cake.

As you can see, the correct consistency or thickness of royal icing depends on its use. Piped designs and string work require fairly thick icing, while flood work requires a thinner product. For this reason, many pastry chefs do not use a recipe for royal icing but prepare small batches as needed, using the Procedure for Preparing Royal Icing, below. For those who prefer working with a recipe, one is provided.

Procedure: Preparing Royal Icing

1. Place desired amount of confectioners' sugar in a mixing bowl. Add a small quantity of cream of tartar (for whiteness), about 1/8 tsp per pound of sugar (0.6 g per kilogram).
2. Beat in egg whites (pasteurized), a little at a time, until the sugar forms a smooth paste. You will need 2 to 3 oz egg whites per pound of sugar (125 g per kilogram).
3. Keep unused icing covered with a damp cloth or plastic film at all times to prevent hardening.

Glazes

Glazes are thin, glossy, transparent coatings that give a shine to baked products and help prevent drying.

The simplest glaze is a sugar syrup or diluted corn syrup brushed while hot onto coffee cakes or Danish pastries. Syrup glazes may also contain gelatin or waxy maize starch.

Fruit glazes for pastries, the most popular of which are apricot and red currant, are available commercially prepared. They are melted, thinned with a little water, syrup, or liquor, and brushed on while hot. Fruit glazes may also be made by melting apricot or other preserves and forcing them through a strainer. It helps to add melted, strained preserves to commercial glazes because these products usually have little flavor.

The glaze recipes included in this chapter are of two types: chocolate and gelatin-based.

Chocolate glazes are usually melted chocolate containing additional fats or liquids, or both. They are applied warm and set up to form a thin, shiny coating.

Gelatin-based glazes, which include many fruit glazes, are usually applied only to the tops of cakes and charlottes made in ring molds.

Rolled Coatings

The three commonly used rolled cake coatings are rolled fondant, marzipan, and modeling chocolate. Rather than being applied by spreading or pouring, these are rolled into thin sheets, using a rolling pin, and draped over the cake to cover it. To ensure the coating adheres to the cake, the cake is first brushed with apricot glaze or a similar product, or iced with a thin layer of buttercream before the rolled coating is applied.

Marzipan is a paste made of ground almonds and sugar.

Rolled fondant is a dough like product consisting primarily of confectioner's sugar combined with small quantities of glucose, water, gelatin, and other ingredients to give it the proper consistency. It is firm and stiff enough to be kneaded, and pliable enough to be rolled out in thin sheets. Like poured fondant, it is almost always purchased ready-prepared.

Modeling chocolate is a stiff paste made of melted chocolate and corn syrup.

5. Frozen Dessert: Ice Cream and Bombes

The popularity of ice cream needs no explanation. Whether it is a plain scoop of vanilla ice cream in a dish or an elaborate assemblage of fruits, syrups, toppings, and numerous flavours of ice cream and sherbet, a frozen dessert appeals to everyone.

Until recently, few establishments made their own ice cream because of the labour involved, the equipment required, and the sanitation regulations and health codes that had to be followed. Furthermore, the wide availability of high-quality commercially produced ice creams made it unnecessary y for operations to prepare their own. But today, many restaurants find that offering their own homemade sorbets and ice creams is appealing to customers. In fact, in the finest restaurants, customers may expect the pastry chef to produce frozen desserts as well as pastries. Thus, learning to make ice cream has become an important skill.

You will find that much of this chapter seems familiar. The base for ice cream, for example, is the same crème anglaise you have used in many other preparations. Other techniques, too, in this chapter, such as preparing syrups and whipping meringues, are used in many areas of the bakeshop.

Churn-Frozen Desserts

ICE CREAM AND sherbet are churn-frozen, meaning they are mixed constantly while being frozen. If they were not churned, they would freeze into solid blocks of ice. The churning keeps the ice crystals small and incorporates air into the dessert.

Ice cream is a smooth, frozen mixture of milk, cream, sugar, flavorings, and, sometimes, eggs. Philadelphia-style ice cream contains no eggs, and French-style ice cream contains egg yolks. The eggs add richness and help make a smoother product because of the emulsifying properties of the yolks.

Ice milk is like ice cream, but with a lower butterfat content. Frozen yogurt contains yogurt in addition to the normal ingredients for ice cream or ice milk.

Sherbets and ices are usually made from fruit juices, water, and sugar. American sherbets usually contain milk or cream and, sometimes, egg whites. The egg whites increase smoothness and volume. Ices, also called water ices, contain only fruit juice, water, sugar, and, sometimes, egg whites; they do not contain milk products. The French word sorbet (pronounced sor BAY) is sometimes used for these products. Granité (pronounced grah nee TAY) is coarse, crystalline ice, made without egg whites.

Italian versions of ice cream, sorbet, and granité are called gelato (plural, gelati), sorbetto (plural, sorbetti), and granita (plural in Italian, granite; or, in common English usage, granitas). Traditional Italian gelato (which means, basically, "frozen") is usually lower in fat than other ice creams. Gelato flavours such as vanilla and chocolate are often made with milk only, and no cream. Fruit gelati often contain cream, but because they are mostly fruit purée, they are still low in fat. In addition, many gelati are made without egg yolks, and most are made without other emulsifiers and stabilizers. Thus, they melt quickly and have a light texture and flavour in the mouth. On the other hand, they are mixed less than ice creams and have low overrun (defined below), contributing to a rich mouth feel (also defined below)

Production and Quality

A basic French- or custard-style ice cream mix is simply a crème anglaise or custard sauce mixed with 1 or 2 parts heavy cream for every 4 parts milk used in the sauce. This base is flavoured, as desired, with vanilla, melted chocolate, instant coffee, sweetened crushed strawberries, and so on, chilled thoroughly, and then frozen according to the instructions for the particular equipment in use.

When the mix has frozen, it is transferred to containers and placed in a deep freeze at below 0°F (–18°C) to harden. (Soft-frozen or soft-serve ice creams and gelati are served directly as they come from the churn freezer, without being hardened.)

Whether you make ice cream or buy it, you should be aware of three quality factors:

1. Smoothness is related to the size of the ice crystals in the product. Ice cream should be frozen rapidly and churned well during freezing so large crystals don't have a chance to form.

   Rapid hardening helps keep crystals small, as do eggs and emulsifiers or stabilizers added to the mix.

   Large crystals may form if ice cream is not stored at a low-enough temperature (below 0°F/–18°C).

2. Overrun is the increase in volume due to incorporation of air when freezing ice cream. It is expressed as a percentage of the original volume of the mix. For example, if the mixture doubles in volume, then the amount of increase is equal to the original volume, and the overrun is 100%.

   Some overrun is necessary to give a smooth, light texture. If ice cream has too much overrun, it is airy and foamy and lacks flavour. It was once thought that ice cream should have from 80 to 100% overrun, and that less would make it heavy and pasty. This may be true for ice creams containing gums and other stabilizers, but some high-quality manufacturers produce rich (and expensive) ice cream with as little as 20% overrun.

   Overrun is affected by many factors, including the type of freezing equipment, thelength of churning time, the fat content of the mix, the percentage of solids in the mix, and how full the freezer is.

3. Mouth feel, or body, depends, in part, on smoothness and overrun, as well as other qualities. Good ice cream melts in the mouth to a smooth, not too heavy liquid. Some ice creams have so many stabilizers they never melt to a liquid. Unfortunately, many people are so accustomed to these products that an ice cream that actually does melt in the mouth strikes them as "not rich enough."

Butterfat from cream contributes to a rich mouth feel. However, too high a fat content can detract from the texture. This is because, when fat content is especially high, some of the fat may congeal into tiny lumps of butter during churn-freezing, producing a grainy texture.

A good gelato has a light, smooth mouth feel, attributable to low fat content and lack of emulsifiers, combined with low overrun.

Storage and Service

Five guidelines are essential to the proper storage and ser vice of churn-frozen desserts:

1. Store ice creams and sherbets below 0°F (–18°C). This low temperature helps prevent the formation of large ice crystals.

2. To prepare for serving, temper frozen desserts at 8° to 15°F (–13° to –9°C) for 24 hours so they will be soft enough to serve.
3. When serving, avoid packing the ice cream. The best method is to draw the scoop across the surface of the product so the product rolls into a ball in the scoop.
4. Use standard scoops for portioning ice cream. Normal portions for popular desserts are as follows:

Parfait	Three No. 30 scoops
Banana split	Three No. 30 scoops
À la mode topping for pie or cake	One No. 20 scoop
Sundae	Two No. 20 scoops
Plain dish of ice cream	One No. 10, 12, or 16 scoops

5. Measure syrups, toppings, and garnishes for portion control. For syrups, use pumps that dispense measured quantities, or use standard ladles.

Popular Ice Cream Desserts

Parfaits are made by alternating layers of ice cream and fruit or syrup in tall, narrow glasses. They are usually named after the syrup or topping. For example, a chocolate parfait has three scoops of vanilla or chocolate ice cream alternating with layers of chocolate syrup, and topped with whipped cream and shaved chocolate. (This is the most common meaning of the term parfait in North America. See page 566 for the original parfait.)

Sundaes or coupes consist of one or two scoops of ice cream or sherbet placed in a dish or glass and topped with any of a number of syrups, fruits, toppings, and garnishes. They are quick to prepare, unlimited in variety, and as simple or as elegant as you wish—served in an ordinary soda fountain glass or in a silver cup or crystal champagne glass.

Coupes are often elegant, attractively decorated desserts. Many types have been handed down from the classic cuisine of years ago. The following are classic coupes and similar desserts that still may be made today, but often with different names (the classical names are given as a point of interest, but with the exception of Peach Melba, Pear Belle Hélène, and Coupe au Marrons, they aren't often used today).

1. **Coupe Arlesienne**: In the bottom of the cup, place a spoonful of diced candied fruits that have been soaked in kirsch. Add a scoop of vanilla ice cream, top with a poached pear half, and coat with apricot sauce.
2. **Coupe Black Forest**: Place a scoop of chocolate ice cream in the cup and add sweet, dark cherries flavored with a little cherry brandy. Decorate with rosettes of whipped cream and shaved chocolate.
3. **Coupe Edna May**: Top vanilla ice cream with sweet cherries. Decorate with whipped cream mixed with enough raspberry purée to color it pink.
4. **Coupe Gressac**: Top vanilla ice cream with three small almond macaroons

moistened with kirsch. Top with a small poached peach half, cut side up, and fill the center of the peach with red currant jelly. Decorate with a border of whipped cream.

5. **Coupe Jacques**: Place a scoop each of lemon sherbet and strawberry ice cream in a cup. Top with a mixture of diced, fresh fruit flavored with kirsch.
6. **Coupe aux Marrons**: Top vanilla ice cream with candied chestnuts (marrons glacés) and whipped cream.
7. **Coupe Orientale**: Place diced pineapple in the bottom of the cup and add pineapple sherbet. Top with apricot sauce and toasted almonds.
8. **Peach Melba**: Top vanilla ice cream with a poached peach half covered with Melba Sauce and topped with slivered almonds.
9. **Pear Belle Hélène**: Top vanilla ice cream with a poached pear half covered with chocolate sauce and garnished with toasted, sliced almonds.

Homemade Ice Creams and Sorbets

The same quality factors that apply to commercially made frozen desserts, described above, apply to those you make yourself.

The first two recipes that follow illustrate the basic procedures for making ice cream and sorbet. Using the procedures in these two recipes, you can make an unlimited variety of frozen desserts, as exemplified by the many variations that follow each main recipe. Following these basic recipes are additional recipes for specialty ice creams and sorbets.

Ice Cream Production

As in other areas in the bakeshop, accurate measurement of ingredients is important. In the case of frozen desserts, proper measurement is important to ensure the mix freezes properly. This is because the ratio of sugar weight to total weight has a strong effect on freezing. If an ice cream or sorbet mix contains too much sugar, it will not freeze enough to become firm. On the other hand, an ice cream with too little sugar will not be as smooth as one with the correct amount.

For a basic vanilla ice cream, the weight of the sugar is usually 16 to 20% of the total weight. Adding other ingredients makes the calculation more complicated because many ingredients, such as fruits, contain sugar. When you are developing new recipes, test a small batch of the mixture to see how hard it freezes, and increase or decrease the quantity of sugar as necessary.

Ice creams have a better texture when the finished mix is refrigerated for about 12 hours before freezing. This maturing time enables the proteins of the eggs and milk to bond with more of the water molecules in the mixture. This bonding leaves fewer water molecules available to form ice crystals, which can give ice cream a grainy texture.

Careful sanitation procedures are critical to ice cream production because the ice cream mix is a good breeding ground for bacteria. Use equipment made of stainless steel or other nonporous and noncorrosive material, and properly clean and sanitize it after every use.

Sorbet Production

A basic sorbet mixture is simply a sugar syrup mixed with flavouring ingredients. For sorbets, the proportion of sugar in the mix is even more crucial to the texture of the final product than it is for ice cream, because sorbets do not contain the cream or egg yolks that contribute to the smooth texture of ice cream. Rather, the size of the ice crystals is the most important factor in texture.

The sugar content of fruit varies with its ripeness and other factors. Therefore, testing the sugar concentration of sorbet mix is the surest way to achieve proper texture. Sugar concentration can be measured with a hydrometer, also called a saccharometer (see photo, top right). Use only the basic measurement of sugar concentration, using the Brix and Baumé scales, for best freezing, sorbet mixtures should be at a concentration of 30° to 32.5° Brix or 16° to 18° Baumé (see photo, lower right). If the sugar density is too high, dilute with a little water. If it is too low, add a little sugar syrup to increase the sugar content.

Rapid freezing produces smaller ice crystals, and thus smoother texture, than slow freezing. For sorbets as well as for ice creams, chill the mixture well before freezing so it freezes in the shortest possible time.

Using a little corn syrup in place of some of the sugar for the syrup may also contribute slightly to smoothness in some sorbets. Classic sorbets, however, are based on a syrup made only with regular granulated sugar. Using corn syrup in a syrup for sorbets darkens it somewhat, because of increased browning of the sugars and starches in the corn syrup. This may be a disadvantage for some white or light-coloured sorbets.

For granitas, unlike for sorbets, large crystals are characteristic of the product and are desirable. Classic granitas are made with sorbet like mixes, but with two differences: First, the sugar content is slightly lower, so the ice crystals are larger. Second, rather than being churn frozen, the mixture is still-frozen in a pan and stirred periodically as it freezes. This freezing method gives granitas their typical icy texture.

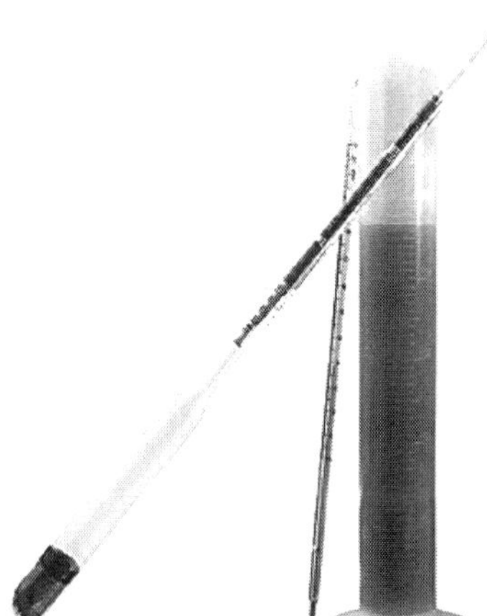

Hydrometer.

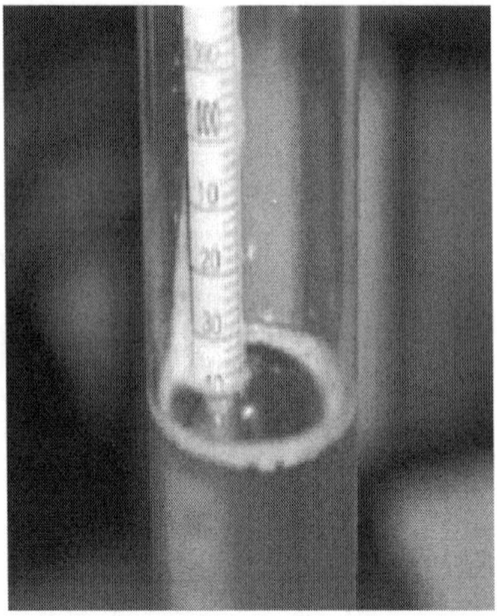

Testing a sorbet mix with a hydrometer.

Vanilla ice-cream among other popular ice cream desserts are Meringues Glacèes and Frozen Éclairs and Profiteroles. Baked Alaska is a popular festive dessert. Although no one is surprised by it anymore, one of the classic names for this dessert is Soufflé Surprise, so called because it looks like baked whipped eggs on the outside but is frozen inside. Following are some popular ice creams:

1. Vanilla bean ice-cream
2. Chocolate ice-cream
3. Cinnamon ice-cream
4. Coffee ice-cream
5. Carob ice-cream
6. Coconut ice-cream
7. Caramel ice-cream
8. Almond, Hazelnut or Macadamia praline ice-cream
9. Cheesecake ice-cream
10. Strawberry ice-cream
11. Raspberry swirl ice-cream
12. Mango ice-cream
13. Peach ice-cream
14. Gingerbread-spice ice-cream
15. Lemon ice-cream
16. Lime ice-cream
17. Honey ice-cream
18. Dulce de leche ice-cream
19. Bitter chocolate ice-cream

Parfaits

- Parfaits are made by alternating layers of ice cream and fruit or syrup in tall, narrow glasses.
- They are usually named after the syrup or topping. For example, a chocolate parfait has three scoops of vanilla or chocolate ice cream alternating with layers of chocolate syrup, and topped with whipped cream and shaved chocolate.
- The mixture for parfaits consists of three elements: a thick, sweet egg yolk foam, an equal volume of whipped cream, and flavorings.

Sundaes or Coupes

Sundaes or coupes consist of one or two scoops of ice cream or sherbet placed in a dish or glass and topped with any of a number of syrups, fruits, toppings, and garnishes.

They are quick to prepare, unlimited in variety, and as simple or as elegant as you wish—served in an ordinary soda fountain glass or in a silver cup or crystal champagne glass.

Coupes are often elegant, attractively decorated desserts.

Many types have been handed down from the classic cuisine of years ago.

Bombes

The bombe is one of the most elegant frozen desserts, often elaborately decorated with fruits, whipped cream, petits fours secs, and other items after unmoulding.

It is made by lining a chilled mould (usually spherical or dome-shaped) with a layer of ice cream or sherbet and freezing it hard.

The centre is filled with a bombe mixture of compatible flavor and then frozen again.

Mixtures for frozen mousses can also be used to fill bombes, as can regular ice cream or sherbet, but a special bombe mixture is the most common choice.

Following are various classic Bombes:

- **Bombe Africaine**

  Coating: chocolate ice cream Filling: apricot bombe mixture
- **BOMBE AIDA**

  Coating: strawberry ice cream Filling: kirsch-flavored bombe mixture
- **BOMBE BRESILIENNE**

  Coating: pineapple sherbet Filling: bombe mixture flavored with vanilla and rum and mixed with diced pineapple
- **BOMBE CARDINALE**

  Coating: raspberry sherbet Filling: praline vanilla bombe mixture
- **BOMBE CEYLON**

  Coating: coffee ice cream Filling: rum-flavored bombe mixture
- **BOMBE COPPELIA**

  Coating: coffee ice cream Filling: praline bombe mixture
- **BOMBE DIPLOMAT**

  Coating: vanilla ice cream Filling: bombe mixture flavored with maraschino liqueur and mixed with candied fruit
- **BOMBE FLORENTINE**

  Coating: raspberry sherbet Filling: praline bombe mixture

- **BOMBE FORMOSA**

  Coating: vanilla ice cream Filling: bombe mixture flavoured with strawberry purée and mixed with whole strawberries

- **BOMBE MOLDAVE**

  Coating: pineapple sherbet Filling: bombe mixture flavoured with orange liqueur

- **BOMBE SULTANE**

  Coating: chocolate ice cream Filling: praline bombe mixture

- **BOMBE TUTTI-FRUTTI**

  Coating: strawberry ice cream or sherbet Filling: lemon bombe mixture mixed with candied fruits

- **CASSATA NAPOLETANA**

- Cassatas are Italian-style bombes lined with three layers of different ice creams and filled with Italian meringue mixed with various ingredients.
- The most popular, Cassata Napoletana, is made as follows:
  1. Line the mould first with vanilla, then with chocolate, and finally with strawberry ice cream.
  2. Fill with Italian Meringue flavoured with vanilla, kirsch, or maraschino and mixed with an equal weight of diced candied fruits. A little whipped cream may be added to the meringue, if desired.

6. Storage of Confectionery Products

Cakes

Cakes are perishable. No single process is the one for properly storing all types of cakes. It is important to know the kind of the cake before deciding how to store it properly. You should learn the proper way of storing the cake in the cake display cabinets.

Cakes are rich desserts. It is baked, and it contains ingredients such as sugar, eggs, flour, and oil. The storage of the cakes should be decided based on the ingredients that it is made up of.

Things that affect the freshness of the cake are heat, humidity, and intense artificial light. The intense light can change the colour of the cake.

Cake can be kept in the best condition if it is stored away from all of these elements.

Storing the cake before Decorating

If the cakes that you are storing has an elaborate decorating plan, then it is better to divide the labor. Cake can be kept fresh if you bake the cake in advance. It will decrease the work as well. It will also prevent the outer layer from becoming stale.

Let the Cake Cool:

1. After baking the do not put the cake in the refrigerated cabinet before it gets cold. If a warm cake is stored and wrapped, it can create steam. The formation of vapor can lead to quick deterioration of the cake. Keep the cake at room temperature until it reaches room temperature. If you are storing is for a longer period, wrap it tightly. You can also display them in the cake display cabinets without wrapping.
2. If you store it in the cake display unit after wrapping it tightly then, it will stay fresh for a few days. The layers of the cake are not used then they can be kept in the freezer for up to a month. Just thaw the layers before using them.

Storing the Cake after Icing

After the cake is iced the methods for cake storage also change. After freezing the storage, the method depends on the type of icing that is used. Here are some of the different types of icing and tips for proper storage:

Fondant and the Buttercream

1. Fondant and the buttercream have the quality of sealing in the moisture. The cake stays fresh, and it does not dry out easily if it has an icing of buttercream and fondant. It creates a barrier between the cake and the outer world.
2. They can be stored in the refrigerated cake display cabinets, but the condensation can cause the colors to fade.
3. To keep the icing color, it is important that you avoid sudden temperature changes. When the cake is taken out of the refrigerator first, transfer it to air conditioned space, so it does not suffer from heat shock.

Whipped Cream or Cream Cheese

1. If you are using whipped cream or cream cheese as icing, then it should not be left at the room temperature for a long time. It should be placed in the refrigerator to maintain its good texture.
2. The cream cheese icing can be frozen. The texture of the frozen cream icing can change.

Boiled Icing

1. The boiled icing can be refrigerated, but it can also be stored at room temperature.
2. Cakes with this icing should not be frozen. If frozen, the texture of this icing can become gummy.

Ganache

1. Cakes with Ganache frosting can be stored in the refrigerator.
2. Storing this cake is similar to storing the cakes with fondant icing.
3. You can store it at room temperature, and it can be frozen as well.

General Tips

1. If the cake that you are making is a tall and elaborate one, then always prefer to store the cake in separate layers. It is convenient and comfortable. Assemble the cake when it needs serving. It will also allow dividing the labor, and you would be able to do a lot of the work beforehand.
2. If the cake that you are decorating a cake that has tinted pieces, then it is possible that it will bleed. The colored pieces of the cake can suffer if it is left open to the moisture. Moisture is not okay for the tinted cake decoration at all.
3. While deciding the storage temperature of the cake, always consider the filling. The room temperature depends on the ingredients that are used in the filling. The filling will decide whether it needs to be stored at room temperature or in the refrigerated cabinets.

Storing the Cake Slices

- If you are storing the cut slices of the cake, then they need to be stored accordingly.
- The slices of the cakes tend to become dry pretty quickly. The freshness of the slices can be maintained by placing a plastic sheet on the piece of cake.
- You can also add some more toppings to seals the exposed side of the slice.

Pastry

- Pastries taste best straight from the oven, gradually losing flavor and appeal as freshness dissipates.
- Oxygen exposure dries out pastries, making them taste stale.
- Without proper packaging, soft pastries become hard and hard pastries become soft.
- Depending on the type of packaging and storage method, you can keep pastries fresh for up to three months.

Room Temperature Storage

- Fresh pastries must cool to room temperature before you wrap them, or the residual heat will create condensation making them soggy.

- Place fresh pastries inside a paper bag to help retain the quality of the crust.
- Keep the pastries at room temperature to prevent condensation so that they don't soften.
- Seal the paper bag inside a plastic storage bag for extended storage of up to about five days.
- Squeeze out any excess air inside the plastic bag to extend freshness

Freezer Storage

- The freezer is the best place to store pastries to retain freshness beyond about five days.
- Use only freezer storage bags designed for optimum moisture-vapor resistance.
- Freeze the pastries for up to two to three months; thaw them in the packaging before reheating.

Serving the Pastries

- Regardless of how you store pastries, pop them in the microwave or oven to help restore freshness.
- Heat pastries at approximately 350 F for 10 minutes in a conventional or toaster oven, or warm them in a microwave on the low setting for 30 to 60 seconds.

Chocolate

- Chocolate is an incredibly popular treat that many people keep as a pantry staple for baking and to eat on its own.
- While all types of chocolate have a pretty long shelf life, it can still spoil if stored improperly, especially over long periods of time.
- Certain conditions, like a heat wave, can cause it to spoil or bloom.

How to Store

- Ideally, chocolate should be stored in a slightly cool, dry, dark place, preferably a consistent 60 to 70 F.
- It's also best for chocolate to be stored somewhere with less than 50 percent humidity to avoid any risk of condensation, as water can cause chocolate to spoil.
- One lesser-known danger is strong odors, as chocolate can absorb the odor and therefore some of the flavors of other foods

How Heat Affects Chocolate

- If your chocolate has melted, this doesn't necessarily mean it's spoiled. While melted chocolate can be messy, it can usually still be saved.
- Once the chocolate cools, it will usually taste the same, even if it's lost its shape.

- However, frequent exposure to high temperatures can cause the cocoa butter in chocolate to rise to the surface, creating bloom, or a cloudy gray layer on the outside.
- While it's safe to eat bloomed chocolate, it doesn't always have the best flavor.
- This is a superficial flaw that can be corrected by melting and properly tempering the chocolate, as long as there aren't any fillings that will get in the way
- In addition to blooming, white chocolate can go rancid when exposed to strong light, which will greatly affect the flavor and quality of the chocolate.

Tips for Warm Climates

- If you live in an extremely warm area and do not have air conditioning, you may need to keep your chocolate in the refrigerator or freezer to prevent it from melting.
- To freeze the chocolate, place it in an airtight container, and do not remove it from its container until it has been brought up to room temperature, which will prevent condensation from forming on chocolate.
- Water droplets will prevent the chocolate from melting smoothly and might affect the texture of the melted chocolate.

Sugar Confectionery

Sugar confectionery items include candies, lollipops, candy bars, chocolate, cotton candy, and other sweet items of snack food. Some of the categories and types of sugar confectionery include the following:

1. Chocolates: Bite-sized confectioneries generally made with chocolate, considered different from a candy bar made of chocolate.
2. Divinity: A nougat-like confectionery based on egg whites with chopped nuts.
3. Dodol: A toffee-like delicacy popular in Indonesia, Malaysia, and the Philippines
4. Dragée: Sugar-coated almonds and other types of sugar panned candies.
5. Fudge: Made by boiling milk & sugar to soft-ball stage. In the US, it tends to be chocolate flavored.
6. Halvah: Confectionery based on tahini, a paste made from ground sesame seeds.
7. Hard candy: Based on sugars cooked to the hard-crack stage. Examples include lollipops, jawbreakers (or gobstoppers), lemon drops, peppermint drops and disks, candy canes, rock candy, etc. Also included are types often mixed with nuts such as brittle

8. Ice cream: Frozen, flavored cream, often containing small pieces of chocolate, fruits and/or nuts.
9. Jelly candies: Including those based on sugar and starch, pectin, gum, or gelatin such as Turkish delight (lokum), jelly beans, gumdrops, jujubes, gummies, etc.
10. Liquorice: Containing extract of the liquorice root, this candy is chewier and more resilient than gums or gelatin candies. For example, Liquorice allsorts. It has a similar taste to star anise.
11. Marshmallow: For example, circus peanuts.
12. Marzipan: An almond-based confection, doughy in consistency.
13. Mithai: A generic term for confectionery in the Indian subcontinent, typically made from dairy products and/or some form of flour. Sugar or molasses are used as sweeteners.
14. Persipan: similar to marzipan, but made with peaches or apricots instead of almonds.
15. Pastillage: A thick sugar paste made with gelatin, water, and confectioner's sugar, similar to gum paste, which is moulded into shapes, which then harden.
16. Tablet: A crumbly milk-based soft and hard candy, based on sugars cooked to the soft ball stage. Comes in several forms, such as wafers and heart shapes. Not to be confused with tableting, a method of candy production.
17. Taffy (British: chews): A sugar confection that is folded many times above 120 °F (50 °C), incorporating air bubbles thus reducing its density and making it opaque.
18. Toffee: A confection made by caramelizing sugar or molasses along with butter. Toffee has a glossy surface and textures ranging from soft and sticky to a hard, brittle material. Its brown color and smoky taste arise from the caramelisation of the sugars.

Storage and Shelf Life

- Shelf life is largely determined by the amount of water present in the candy and the storage conditions.
- High-sugar candies, such as boiled candies, can have a shelf life of many years if kept covered in a dry environment.
- Spoilage of low-moisture candies tends to involve a loss of shape, color, texture, and flavor, rather than the growth of dangerous microbes.
- Impermeable packaging can reduce spoilage due to storage conditions.
- Candies spoil more quickly if they have different amounts of water in different parts of the candy (for example, a candy that combines marshmallow and nougat), or if they are stored in high-moisture environments. This process is due to the effects of water activity, which results in the transfer of unwanted water from a high-moisture environment into a low-moisture candy, rendering it rubbery, or the loss

of desirable water from a high-moisture candy into a dry environment, rendering the candy dry and brittle.

- Another factor, affecting only non-crystalline amorphous candies, is the glass transition process. This can cause amorphous candies to lose their intended texture.

Biscuits and Cookies

How to store biscuits

- Keep freshly baked biscuits at room temperature for 1 or 2 days.
- You'll need to cover them with foil or cling film in a plastic bag to prevent drying out.
- If you want to keep them in the fridge, they'll keep well for about 1 week when properly stored.
- Don't serve biscuits that have an odd smell or appearance.
- Freezing biscuits properly:
- To freeze biscuits, wrap them tightly with cling film or foil or place into a freezer bag. They'll then keep for about 2 to 3 months.

The best way to store cookies

- Store cookies at room temperature
- To keep cookies at their best, place greaseproof paper between the cookies and keep in an airtight container for 1 week.
- If your cookies have gone a little stale, restore them to their former chewiness by sprinkling a few drops of water over the cookies on a baking tray and place in a hot oven for a few minutes.

How to freeze cookie dough

- You can store uncooked cookie dough by wrapping it in cling film and storing in the fridge for up to 2 days.
- Alternatively, cookie dough will store for up to 4 months in the freezer. You can either wrap a whole dough ball in cling film and then place in the freezer.
- To make life easier, you can roll the dough into a log and wrap it with cling film before freezing.
- When you are ready to bake, use a hot knife to slice the dough into rounds straight from the freezer.
- Increase the baking time of the cookies by a few extra minutes and keep an eye on them to compensate for the lower temperature of the dough.

Ice Cream

- Ice cream is a perishable product and should be treated carefully.
- When frozen desserts are exposed to temperatures above 10°F, they become subject to adverse changes in body, texture and flavor characteristics.

- Although individual manufacturers' recipes yield ice cream of varying consistency and flavor, all ice cream will be negatively affected if improperly handled or stored.

Here are some tips on how to keep ice cream in its most delectable form:

In the Store

- Make the ice cream aisle your last stop during your trip to the supermarket.
- Check the temperature of your grocer's freezer case. The optimum temperature is 0°F (-18°C) or colder. The temperature in the supermarket's freezer case should not be above 10°F (-12°C).
- If kept at a proper temperature, ice cream will be thoroughly frozen and will feel hard to the touch. If the product is soft, you may wish to bring it to the attention of the store manager.
- In an open top freezer case, always select ice cream and frozen treats stored below the freezer line.
- Put ice cream products in the separate section of your grocery cart, or place on top of other groceries.
- Insulate ice cream products for the ride home. When your groceries are packed, request a freezer bag or additional brown paper bag to insulate your ice cream.
- Make the grocery store or ice cream parlor your last errand before going home. This will insure that your ice cream does not sit in a warm car while you are making other stops.

At Home

- Do not allow ice cream to repeatedly soften and re-freeze. When ice cream's small ice crystals melt and re-freeze, they can eventually turn into large, unpalatable lumps.
- Your freezer should be set at between -5°F and 0°F. Ice cream is easy to dip between 6°F and 10°F, the ideal serving temperature range.
- Store ice cream in the main part of the freezer. Do not store ice cream in the freezer door, where ice cream can be subject to more fluctuating temperatures since the door is repeatedly open and shut.
- Keep the ice cream container lid tightly closed when storing in the freezer to reduce the formation of ice crystals. Once you have opened the container, you can press a piece of plastic wrap or waxed paper directly on the surface of the ice cream before replacing the lid. Put it back in the freezer immediately after scooping. For long periods of storage, putting the ice cream in a gallon freezer bag will also help minimize ice crystals.
- Do not store ice cream alongside uncovered foods; odors may penetrate ice cream and affect its flavour.

Glossary

Many culinary terms in common use are taken from the French. Phonetic guides for difficult-to-pronounce words are included here, using English sounds. However, exact renderings are impossible in many cases because the French language has a number of sounds that do not exist in English.

1-2-3 dough	A pastry dough made of one-part sugar, two parts fat, and three parts flour, by weight.
Absorption	The amount of water a flour can take up and hold while being made into a simple dough. Absorption is based on a predetermined standard dough consistency or stiffness; expressed as a percentage of the weight of flour.
Active dry yeast	A dry, granular form of yeast that must normally be rehydrated before using.
Aerobic	Requiring oxygen to live and grow; said of some bacteria.
Air cell	A tiny bubble of air, created by creaming or foaming, that assists in leavening a dough or batter.
Allergen	A substance that triggers an allergic reaction.
All-purpose flour	Flour formulated to be slightly weaker than bread flour so it also can be used for pastries.
Allumette	French for "matchstick"; any puff pastry item made in thin sticks or strips.
Almond paste	A mixture of finely ground almonds and sugar.
Amylase	An enzyme in flour that breaks down starches into simple sugars.
Anaerobic	Requiring an absence of oxygen to live and grow; said of some bacteria.
Anaphylaxis	A sudden and severe allergic reaction of the immune system.

Angel food cake	A type of cake made of meringue (egg whites and sugar) and flour.
Angel food method	A cake-mixing method that involves folding a mixture of flour and sugar into a meringue.
Apple charlotte	A dessert of apples cut up and baked in a mold lined with bread slices.
AP weight	As purchased; the weight of an item before trimming.
Artisan bread	Bread made by a skilled manual worker; usually refers to handmade breads made using traditional methods and with natural ingredients only.
Ash	The mineral content of flour; expressed as a percentage of the total weight.
Autolyse	A resting period early in the mixing procedure of yeast doughs during which the flour fully absorbs the water.
Baba	A type of yeast bread or cake that is soaked in syrup.
Babka	A type of sweet yeast bread or coffee cake.
Bagel	A ring-shaped lean yeast dough product made from a very stiff dough.
Bagged	A cookie makeup method in which the dough is shaped and deposited on the pan or sheet using a pastry bag.
Baked Alaska	A dessert consisting of ice cream on a sponge cake base, covered with meringue and browned in the oven.
Baked custard	A custard that is baked undisturbed so it sets into a solid.
Baked meringue	Any meringue mixture that is baked until dry Baker's cheese : A soft, unaged cheese used to make pastry fillings, cheesecake, and similar products.
Baker's percentage	A method of expressing ratios of ingredients in a baking formula in which the weight of each ingredient is expressed as a percentage of the weight of the flour.
Baking ammonia	A leavening ingredient that releases ammonia gas and carbon dioxide.
Baking chocolate	A chocolate product in which another fat is substituted for part of the cocoa butter.
Baklava	A Greek or Middle Eastern dessert made of nuts and phyllo dough and soaked with syrup.

Bar	A cookie makeup method in which the dough is shaped into flattened cylinders, baked, and sliced crosswise into individual cookies; also, a cookie made by this method.
Barm	A thin or batter-like sourdough starter.
Batter	A semiliquid mixture containing flour or other starch, used for the production of such products as cakes and breads and for coating products to be deep fried.
Baumé	A unit of measure indicating the specific gravity of a solution, often used to indicate sugar concentration.
Baumkuchen (BOWM koo khen)	A cake made by adding one thin layer of batter at a time to a pan and browning lightly under a broiler after each addition, repeating until the cake is the desired thickness.
Bavarian cream	A light, cold dessert made of gelatine, whipped cream, and custard sauce or fruit.
Bavarois (bah var WAH)	French for Bavarian cream.
Beignet soufflé (ben YAY soo FLAY)	A type of fritter made with éclair paste, which puffs up greatly when fried.
Benching	An intermediate fermentation and resting period for yeast doughs, after folding and before rounding.
Betty	A baked dessert consisting of layers of fruit and cake crumbs.
Biga	A yeast pre-ferment made as a stiff dough.
Biscuit method	A mixing method in which the fat is mixed with the dry ingredients before the liquid ingredients are added.
Black Forest torte	A chocolate sponge layer cake filled with whipped cream and cherries.
Blancmange (blaw MAWNGE)	(1) An English pudding made of milk, sugar, and corn starch. (2) A French dessert made of milk, cream, almonds, and gelatine.
Blitz puff pastry	A type of pastry mixed like a very flaky pie dough, then rolled and folded like puff pastry.
Bloom	(1) A whitish coating on chocolate caused by separated cocoa butter. (2) To hydrate gelatine. (3) The relative strength or gelling power of a grade of gelatine.
Blown sugar	Pulled sugar made into thin-walled, hollow shapes by being blown up like a balloon.
Boiled icing	Italian meringue used as a cake icing.

Bolting	The process of sifting flour, primarily to separate the bran.
Bombe	A type of frozen dessert made in a dome-shaped mould.
Boston cream pie	Not a pie, but a sponge cake or other yellow cake filled with pastry cream and topped with chocolate fondant or confectioners' sugar. Boulanger (boo lawn ZHAY) : The bread baker, who prepares breads and other yeast goods, including such breakfast items as brioche, croissants, and Danish pastry.
Boulanger, A.	An eighteenth-century Parisian credited with starting the first restaurant.
Bran	The hard outer covering of kernels of wheat and other grains.
Bran flour	Flour to which bran flakes have been added.
Bread flour	Strong flour, such as patent flour, used for breads.
Break system	A milling system to produce various grades of flour by repeatedly breaking the grains between rollers, and sifting.
Brioche	Rich yeast dough containing large amounts of eggs and butter; a product made from this dough.
Brix	A unit of measure indicating the sugar concentration of a solution.
Brown sugar	Regular granulated sucrose containing impurities that give it a distinctive flavour and colour.
Buttercream	An icing made of butter and/or shortening blended with confectioners' sugar or sugar syrup and, sometimes, other ingredients.
Butterfat	The fat in dairy products. Also called milk fat.
Buttermilk	(1) The milky liquid drained off after cream is churned to make butter. Rarely sold. (2) Milk, usually low-fat or fat-free, that has been cultured by bacteria to resemble the original buttermilk (definition 1).
Cabinet pudding	A baked custard containing sponge cake and fruit.
Cake flour	A fine white flour made from soft wheat.
Calorie	The amount of heat needed to raise the temperature of 1 kilogram of water 1 degree Celsius.
Cannoli	Fried Italian pastries made in tube shapes, generally with a sweet cream or cheese filling (singular form : cannolo).

Caramelization	The browning of sugars caused by heat.
Caramelize	To change sugar into caramel by means of heat.
Carbohydrate	Any of a group of compounds, including starches and sugars, that supply energy to the body.
Carême, Marie-Antoine	Important and influential nineteenth-century cook and pastry chef.
Carotenoid	An orange-yellow pigments present in many plant products, including unbleached flour; responsible for the creamy colour of flour.
Cassata	An Italian-style bombe, usually with three layers of different ice creams, plus a filling of Italian meringue.
Cast sugar	Sugar boiled to the hard-crack stage and then poured into moulds to harden. Also called poured sugar.
Celiac disease	A reaction to gluten in which the lining of the intestine is damaged.
Celsius scale	The metric system of temperature measurement, with 0°C at the freezing point of water and 100°C at the boiling point of water.
Centi-	Prefix in the metric system meaning one-hundredth.
Challah	A rich egg bread, often made as a braided loaf.
Charlotte	(1) A cold dessert made of Bavarian cream or other cream in a special mould, usually lined with ladyfingers or other sponge products. (2) A hot dessert made of cooked fruit and baked in a special mould lined with strips of bread.
Charlotte ring	A metal ring used as a mould for charlottes and other desserts.
Chef de cuisine	The head of a kitchen.
Chef garde manger	Pantry chef.
Chemical leavener	A leavener such as baking soda, baking powder, or baking ammonia, which releases gases produced by chemical reactions.
Chiffon cake	A light cake made by the chiffon method.
Chiffon method	A cake-mixing method that involves the folding of whipped egg whites into a batter made of flour, egg Chocolate : Any of a number of products made from fermented, roasted, yolks, and oil.
Chiffon pie	A pie with a light, fluffy filling containing egg whites and, usually, gelatine. ground cocoa (or cacao) beans. Often with the addition of sugar, flavourings, and other ingredients.

Chocolate liquor	Unsweetened chocolate, consisting of cocoa solids and cocoa butter. Also called cocoa mass.
Chocolate truffle	A small ball of chocolate ganache, served as a confection.
Cholesterol	A fatty substance found in foods derived from animal products and in the human body; a high level of cholesterol has been linked to heart disease.
Christmas pudding	A dark, heavy steamed pudding made of dried and candied fruits, spices, beef suet, and crumbs.
Ciabatta	A type of Italian bread made from a very slack dough deposited on pans with minimal shaping.
Clean up stage	A stage of yeast dough mixing in which the ingredients become fully incorporated into a dough; so-called because the formed dough "cleans up" formerly unmixed ingredients from the mixing bowl.
Clear flour	A tan-colored wheat flour made from the outer portion of the endosperm.
Coagulation	The process by which proteins become firm, usually when heated.
Coating chocolate	A sweetened chocolate similar in appearance to couverture but with other fats substituted for part of the cocoa butter.
Cobbler	A fruit dessert similar to a pie, but without a bottom crust.
Cocoa	The dry powder that remains after cocoa butter is pressed out of chocolate liquor.
Cocoa bean	Seed of the cacao tree. Fermented, roasted, and ground to make cocoa and chocolate products.
Cocoa butter	A white or yellowish fat found in natural chocolate.
Cocoa mass	Unsweetened chocolate, consisting of cocoa solids and cocoa butter. Also called chocolate liquor.
Cold charlotte	A dessert consisting of Bavarian cream made in a mould lined with a sponge-cake product.
Common meringue	Egg whites and sugar whipped to a foam. Also called French meringue.
Complex presentation	A dessert plating style consisting of an arrangement of two or more desserts plus sauces and garnishes.
Complex sugar	A large sugar molecule containing at least 12 carbon atoms. Sucrose or table sugar is a complex sugar. See also Simple sugar.

Compote	Cooked fruit served in its cooking liquid, usually a sugar syrup.
Compressed yeast	Live, moist yeast, made into dense cakes. Also called Fresh yeast.
Conching	A step in the manufacturing of chocolate, the purpose of which is to create a fine, smooth texture.
Condensed milk	Heavily sweetened milk that has had 60 percent of the water content removed.
Confectioners' chocolate	See Couverture.
Confectioners' sugar	Sucrose ground to a fine powder and mixed with a little corn starch to prevent caking.
Confisseur (cone fee SUR)	A confectioner, or candy maker.
Contact method	A decorating technique in which the tip of a paper cone of icing stays in contact with the decorated surface.
Contaminated	Containing a harmful substance not originally present in the food.
Cooked fruit method	A method for making pie fillings in which the fruit is cooked and thickened before being placed in the pie crust.
Cooked juice method	A method for making pie fillings in which the fruit juices are cooked, thickened, and mixed with the fruit.
Cookie	North American name for a small, flat, baked treat, usually containing fat, flour, eggs, and sugar. Known in England and other English-speaking countries as "biscuit."
Corn starch pudding	A sweetened liquid, usually milk and flavourings, boiled with corn starch to thicken it.
Corn syrup	A syrup made from corn, consisting mostly of glucose. Corrective action : In the HACCP system, a procedure that must be followed whenever a critical limit is not met.
Coulis (koo LEE)	A sweetened fruit purée, used as a sauce.
Coupe	A dessert consisting of one or two scoops of ice cream or sherbet placed in a dish or glass and topped with any of a number of syrups, fruits, toppings, and garnishes; a sundae.
Couverture	Natural, sweet chocolate containing no added fats other than natural cocoa butter; used for dipping, moulding, coating, and similar purposes. Also called Confectioner's chocolate.

Cracked wheat	A type of wheat meal in which the grains are broken into coarse pieces.
Cream cheese	A soft, fresh cheese with a high milk fat content.
Cream pie	An unbaked pie containing a pastry cream-type filling.
Cream pudding	A boiled pudding made of milk, sugar, eggs, and starch.
Creaming	The process of beating fat and sugar together to blend them uniformly and to incorporate air.
Creaming method	A mixing method that begins with the blending of fat and sugar; used for cakes, cookies, and similar items.
Crème anglaise (krem awn GLEZZ)	A light vanilla-flavoured custard sauce made of milk, sugar, and egg yolks.
Crème brûlée	French for "burnt cream"; a rich custard with a brittle top crust of caramelized sugar.
Crème caramel	A custard baked in a mold lined with caramelized sugar, then unmoulded.
Crème chantilly (krem shawn tee YEE)	Sweetened whipped cream flavoured with vanilla.
Crème Chiboust (krem shee BOO)	A cream filling made of pastry cream, gelatine, meringue, and flavourings.
Crème fraîche (krem fresh)	A slightly aged, cultured heavy cream with a slightly tangy flavour.
Crémeux	A type of cream or pudding consisting of crème anglaise plus one or more thickeners or binders, such as chocolate, gelatin, or butter.
Crêpe (krep)	A very thin French pancake, often served rolled around a filling.
Crêpes Suzette	French pancakes served in a sweet sauce flavoured with orange.
Crisp	(1) A baked fruit dessert with a streusel topping. (2) A confection or garnish consisting of a very thin slice of fruit that has been dried.
Critical control point (CCP)	An action that can be taken to eliminate or minimize a food safety hazard.
Croissant (krwa SAWN)	A flaky, buttery yeast roll shaped like a crescent and made from a rolled-in dough.
Cross-contamination	The transfer of pathogens to food from another food or from work surfaces or equipment.
Crumb	The interior of a baked item, distinct from the crust.

Crumb crust	A piecrust made of cookie crumbs, butter, and sugar.
Crystallize	To form crystals, as in the case of dissolved sugar.
Cuisinier (kwee zeen YAY)	A cook; the head of a kitchen.
Custard	A liquid thickened or set by the coagulation of egg protein.
Dark chocolate	Sweetened chocolate that consists of chocolate liquor and sugar.
Dark couverture	Couverture consisting of chocolate liquor and sugar; contains no milk solids.
Deci-	Prefix in the metric system meaning one-tenth.
Décor	Small food items whose primary purpose is decoration.
Décorateur	A pastry chef who specializes or is skilled in decorative work, such as showpieces, sugar work, and fancy cakes.
Degree Celsius	Unit of measure of temperature in the metric system. One degree Celsius is 1/100 of the temperature range between the freezing point and the boiling point of water.
Demerara sugar	A type of crystalline brown sucrose.
Dessert syrup	A flavoured sugar syrup used to flavour and moisten cakes and other desserts.
Devil's food cake	A chocolate cake made with a high percentage of baking soda, which gives the cake a reddish colour.
Diastase	Various enzymes found in flour and in diastatic malt that convert starch into sugar.
Disaccharide	A complex or double sugar, such as sucrose.
Dobos torte	A Hungarian cake made of seven thin layers, filled with chocolate buttercream, and topped with caramelized sugar.
Docking	Piercing or perforating pastry dough before baking in order to allow steam to escape and to avoid blistering.
Double-acting baking powder	Baking powder that releases some of its gases when it is mixed with water and the remaining gases when it is heated.
Double-panning	Placing a baking sheet or pan on or in a second pan to prevent scorching the bottom of the product being baked.
Dough conditioner	Any of a variety of ingredients added by the baker during production of yeast products to improve gluten development, aid yeast fermentation, and delay staling. Also called dough improver.

Dough relaxation	A period of rest in the production of yeast dough during which gluten strands become adjusted to their new length and become less tight.
Dough strength	An indication of the texture and gluten development of a yeast dough; a combination of elasticity, tenacity, and extensibility.
Drained weight	The weight of solid canned fruit after draining off the juice.
Dredge	To sprinkle or coat thoroughly with sugar or another dry powder.
Dried whole milk	A powdered form of whole milk with the water content removed.
Drop batter	A batter that is too thick to pour but will drop from a spoon in lumps.
Dropped	A cookie makeup method in which portions of dough are measured with a scoop or spoon and dropped onto a baking pan.
Drop-string method	A decorating technique in which the tip of a paper cone of icing stays above the decorated surface and the icing drops as a string onto the surface. Also used to suspend strings of icing between two points.
Durum flour	Flour made from durum wheat, a high-gluten wheat, and used primarily to make spaghetti and other dried pasta.
Dutch process cocoa	Cocoa processed with an alkali to reduce its acidity. Also called Dutched cocoa.
Éclair	A cylindrical piece of baked éclair paste with a pastry cream filling.
Éclair paste	A paste or dough made of boiling water or milk, butter, flour, and eggs; used to make éclairs, cream puffs, and similar products.
Egg-foam cake	A cake leavened primarily by whipped eggs; it usually has a low percentage of fat.
Elasticity	The ability of a dough to spring back when it is stretched.
Empty calorie	A food that provides few nutrients per calorie.
Emulsified shortening	Shortening containing emulsifiers and used for high-ratio cakes.
Emulsion	A uniform mixture of two or more normally unmix able substances.

Endosperm	The starchy inner portion of grain kernels.
English muffin	A yeast dough product made in the shape of a disk and cooked on a griddle.
Enriched flour	Flour to which vitamins and minerals are added to compensate for the nutrients lost when the bran and germ are removed.
EP weight	Edible portion; the weight of an item after trimming.
Escoffier, Georges-August	Most important chef of late nineteenth and early twentieth century; organized cooking principles and kitchen hierarchy.
Evaporated milk	Milk, either whole or skim, that has had 60 percent of the water removed. Extensibility : The ability of a dough to be stretched.
Extract	A flavouring ingredient consisting of flavourful oils or other substances dissolved in alcohol.
Extraction	The portion of the grain kernel separated into a particular grade of flour. Usually expressed as a percentage.
Facultative	Able to live and grow with or without the presence of oxygen; said of some bacteria.
Fat	Any of a group of compounds consisting of chains of fatty acids that supply energy to the body in a concentrated form.
Fermentation	The process by which yeast changes carbohydrates into carbon dioxide gas and alcohol.
Fibre	A type of complex carbohydrate that is not absorbed by the body but is necessary for the proper functioning of the digestive system.
Final development stage	The stage of yeast dough mixing in which gluten becomes smooth and elastic.
Flaky piecrust	A piecrust that has a flaky texture due to layers of fat sandwiched between layers of dough.
Flaky pie dough	A pie dough that has a flaky texture when baked. See Flaky piecrust.
Flat icing	A simple icing made of confectioners' sugar and water, usually used for Danish pastries and sweet rolls.
Flooding	Covering a plate, or a portion of a plate, with sauce
Flour-batter method	A cake-mixing method in which the flour is first mixed with the fat.

Flow of food	The path food travels in a food service operation, from receiving to serving.
Foaming	The process of whipping eggs, with or without sugar, to incorporate air.
Focaccia	A flat Italian bread similar to a thick pizza dough.
Fondant	A type of icing made of boiled sugar syrup that is agitated so it crystallizes into a mass of extremely small white crystals.
Food Danger Zone	The temperature range of 41° to 135°F (5° to 57°C), in which bacteria grow rapidly.
Food intolerance	A nonallergic reaction to a food that may be characterized by any of a variety of undesirable symptoms.
Formula	A set of ingredients and quantities and, usually, instructions for preparing a bakery product; a baker's recipe.
Fortified nonfat or low-fat milk	Milk that has had all or part of the fat removed and that contains added substances, such as vitamins A and D, that increase its nutritional value.
Fougasse	A regional French bread made in the shape of a trellis or ladder.
Four-fold	A technique used to increase the number of layers in puff pastry or Danish pastry by folding the dough in fourths.
Frangipane	A type of almond-flavoured cream.
French bread	Any of a variety of crisp-crusted yeast breads usually consisting only of flour, water, yeast, and salt.
French doughnut	A fried pastry made of choux paste.
French meringue	Egg whites and sugar whipped to a foam; also called common meringue.
French pastry	Any of a variety of small fancy cakes and other pastries, usually in single-portion sizes.
French-style ice cream	Ice cream containing egg yolks.
Fresh yeast	See Compressed yeast.
Fritter	A deep-fried item made of or coated with a batter or dough.
Frozen mousse	A still-frozen dessert containing whipped cream.
Frozen soufflé	A frozen mousse served in a soufflé dish or ramekin so that it resembles a baked soufflé.

Frozen yogurt	A frozen dessert similar to ice cream but made with yogurt instead of or in addition to milk.
Fruit cake	A loaf cake containing a high percentage of dried and candied fruits and, usually, nuts.
Fruit gratin	A dessert consisting of fruit plus a topping, browned under a broiler.
Fruit pie	A baked single- or double-crust pie with a fruit filling.
Fruit torte	A layer cake topped with a decorative arrangement of fruit.
Fungus	A class of organisms that includes yeasts, moulds, and mushrooms.
Fusion cuisine	The use of techniques and ingredients from more than one regional cuisine in a single dish.
Ganache (gah NAHSH)	A rich cream made of sweet chocolate and heavy cream.
Garnish	An edible item added to another food as a decoration or accompaniment.
Gâteau (gah tow)	French word for "cake."
Gâteau St-Honoré	A pastry consisting of a base made of short pastry and pâte à choux and a cream filling, usually crème Chiboust or crème diplomat.
Gaufre (GO fr)	French for "waffle."
Gelatine	A water-soluble protein extracted from animal tissue; used as a jellying agent.
Gelatinization	The process by which starch granules absorb water and swell in size.
Gelato	Italian ice cream.
Gelée (zhuh LAY)	(1) A liquid thickened with gelatine. (2) Any other kind of jelly, especially one set with pectin.
Genoise	A sponge cake made by whipping whole eggs with sugar and folding in flour and, sometimes, melted butter.
Germ	The plant embryo portion of a grain kernel.
Glacé (glah SAY)	(1) Glazed; coated with icing; (2) frozen.
Glacier (glah see YAY)	A chef whose specialty is ice cream.
Glaze	(1) n. A shiny coating, such as a syrup, applied to a food. (2) v. To make a food shiny or glossy by coating it with a glaze or by browning it under a broiler or in a hot oven.

Gliadin	A protein in wheat flour that combines with another protein, glutenin, to form gluten.
Glucose	A simple sugar available in the form of a clear, colourless, tasteless syrup.
Gluten	An elastic substance formed from proteins present in wheat flours that gives structure and strength to baked goods.
Glutenin	See Gliadin.
Gluten window	A thin membrane of yeast dough made in order to test gluten development.
Gram	The basic unit of weight in the metric system; equal to about one thirtieth of an ounce.
Granita	Italian equivalent of the French term Granité.
Granité (grah nee TAY)	A coarse, crystalline frozen dessert made of water, sugar, and fruit juice or another flavouring.
Granulated sugar	Sucrose in a fine crystalline form.
Gum paste	A type of sugar paste or pastillage made with vegetable gum.
HACCP	Hazard Analysis Critical Control Points. HACCP is a food safety system of self-inspection designed to highlight hazardous foods and to ensure proper food handling, by identifying, monitoring, and controlling dangers of food contamination.
Half-and-half	A kind of high-fat milk or low-fat cream containing 10 to 18 percent milk fat.
Hard meringue	A meringue baked until crisp.
Hard sauce	A flavored mixture of confectioners' sugar and butter; often served with steamed puddings.
Hard wheat	Wheat high in protein.
Hazard	A potentially dangerous food condition caused by contamination, growth of pathogens, survival of pathogens, or presence of toxins.
Head baker	The professional in charge of staff and production in a bakery.
Hearth bread	A bread baked directly on the bottom of the oven, not in a pan.
Heavy pack	A type of canned fruit or vegetable with very little added water or juice High-fat cake : A cake with a high percentage of fat; distinguished from a sponge or egg-foam cake.

High-ratio method	See Two-stage method.
High-ratio	(1) Term referring to cakes and cake formulas mixed by a special method and containing more sugar than flour. (2) The mixing method used for these cakes. (3) Term referring to certain specially formulated ingredients used in these cakes, such as shortening.
Homogenized milk	Milk processed so the cream does not separate out.
Hot milk and butter sponge	A sponge cake batter in which a mixture of warm milk and melted butter is mixed into the batter.
Hydration	The process of absorbing water.
Hydrogenation	A process that converts liquid oils to solid fats (shortenings) by chemically bonding hydrogen to the fat molecules.
Ice cream	A churn-frozen mixture of milk, cream, sugar, flavourings, and, sometimes, eggs.
Ice milk	A frozen dessert similar to ice cream, but with a lower fat content.
Icebox	A cookie makeup method in which the dough is shaped into cylinders, refrigerated until firm, and then sliced.
Ice	A frozen dessert made of water, sugar, and fruit juice.
Icing comb	A plastic triangle with toothed or serrated edges; used for texturing icings.
Icing screen	A screen on which cakes are set when iced or glazed, allowing excess to run off.
Improved mix	A yeast dough mixing technique that combines a medium mixing period with a medium fermentation period.
Initial development stage	The first part of the development stage of mixing yeast doughs, in which the dough still appears rough and undermixed.
Instant dry yeast	A dry, granular yeast product that does not require hydration before being added to doughs.
Instant starch	A starch that thickens a liquid without cooking because it has been precooked.
Intensive mix	A yeast dough mixing technique that combines a long mixing period with a short fermentation period.
Inversion	A chemical process in which a double sugar splits into two simple sugars.

Invert sugar	A mixture of two simple sugars, dextrose and levulose, resulting from the breakdown of sucrose.
Isomalt	A compound derived from sucrose, used as a dietary sugar substitute and as an alternative to sugar in decorative sugar work.
Italian meringue	A meringue made by whipping a boiling syrup into egg whites.
Jalebi	A type of Indian dessert made of deep-fried batter soaked in syrup.
Japonaise (zhah po nez)	A baked meringue flavoured with nuts.
Kernel paste	A nut paste, similar to almond paste, made of apricot kernels and sugar.
Kilo-	Prefix in the metric system meaning one thousand.
Kirsch	A clear alcoholic beverage distilled from cherries.
Kirsch torte	A torte made of genoise, meringue disks, and buttercream, and flavored with kirsch.
Kugelhopf	A type of rich, sweet bread or coffee cake, usually made in a tube-type pan.
Lactobacilli	A group of bacteria primarily responsible for producing the acidity in sourdough starters.
Lacto-ovo-vegetarian	A vegetarian diet that allows dairy products and eggs.
Lactose	A form of sugar naturally present in milk.
Lacto-vegetarian	A vegetarian diet that allows milk and other dairy products.
Ladyfinger	A small, dry, finger-shaped sponge cake or cookie.
Lag phase	A period after bacteria have been introduced to a new environment and before they begin to grow and reproduce.
Laminated dough	A dough consisting of many alternating layers of dough and butter or other fat.
Langue de chat (lawng duh SHAH)	A thin, crisp cookie. The French name means "cat's tongue," referring to the shape of the cookie.
Lattice crust	A top crust for a pie made of strips of pastry in a crisscross pattern.
Lean dough	A dough low in fat and sugar.
Leavening	The production or incorporation of gases in a baked product to increase volume and to produce shape and texture.
Lecithin	An emulsifier usually derived from soybeans.

Levain	Sourdough starter.
Levain-levure	French for yeast pre-ferment.
Levure	Commercial yeast.
Light cream	Cream with a fat content of 18 to 30%.
Linzertorte	A tart made of raspberry jam and a short dough containing nuts and spices.
Lipid	Any of a group of compounds containing fats and cholesterol.
Liquid levain	A thin or batter-like sourdough starter.
Liter	The basic unit of volume in the metric system; equal to slightly more than 1 quart.
Long-fermentation dough	A yeast dough that requires a long fermentation period.
Low-fat milk	Milk with a fat content of 0.5 to 2%.
Macaron	French spelling for macaroon.
Macaroon	A cookie made of eggs (usually whites) and almond powder, almond paste, or coconut.
Maillard reaction	A chemical reaction that causes the browning of proteins and sugars together when subjected to heat.
Malt syrup	A type of syrup containing maltose sugar, extracted from sprouted barley.
Marble	To partly mix two colours of cake batter or icing so the colours are in decorative swirls.
Margarine	An artificial butter product made of hydrogenated fats and flavourings.
Marron	French for "chestnut."
Marshmallow	A light confection, icing, or filling made of meringue and gelatine (or other stabilizers).
Marshmallow icing	Boiled icing with the addition of gelatine.
Marzipan	A paste or confection made of almonds and sugar and often used for decorative work.
Mature (dough)	The ideal stage of development for a yeast dough.
Mature (fruit)	Fruit that has completed its development and is physiologically capable of continuing the ripening process, even after removal from the plant.
Maza	An early type of bread; cakes of grain paste baked by the ancient Greeks.
Meal	Coarsely ground grain.

Mealy piecrust	A piecrust in which the fat has been mixed in thoroughly enough so the dough does not have a flaky texture.
Mealy pie dough	A pie dough that has a mealy texture when baked.
Melba sauce	A sweet sauce made of puréed raspberries and, sometimes, red currants.
Meringue	A thick, white foam made of whipped egg whites and sugar.
Meringue chantilly (shawn tee YEE)	Baked meringue filled with whipped cream.
Meringue glacée	Baked meringue filled with ice cream.
Meter	The basic unit of length in the metric system; slightly longer than 1 yard.
Metric system	A measurement system based entirely on decimals.
Microorganism	A life form, such as bacteria, too small to be seen without a microscope.
Milk chocolate	Sweetened chocolate containing milk solids.
Milk chocolate couverture	Couverture consisting of chocolate liquor, sugar, and milk solids. Milk fat : The fat content of milk; also called butterfat.
Millefeuille (mee foy)	French term for napoleon; literally, "thousand leaves." Also used for various layered desserts.
Milli-	Prefix in the metric system meaning one-thousandth.
Mineral	An inorganic element, such as calcium, iron, potassium, sodium, or zinc, that is essential to nutrition
Mixed fermentation	A type of yeast dough fermentation utilizing both a pre-ferment or sour starter and a commercial yeast.
Modeling chocolate	A thick paste made of chocolate and glucose that can be moulded by hand into decorative shapes.
Modified straight dough method	A mixing method similar to the straight dough method, except the fat and sugar are mixed together first to ensure uniform distribution. Used for rich doughs.
Molasses	A heavy brown syrup made from sugarcane.
Moulded	A cookie makeup method in which the dough is shaped into cylinders, cut into equal portions, and shaped as desired.
Monosaccharide	A simple or single sugar such as glucose and fructose.

Monounsaturated fat	A type of fat, normally liquid at room temperature, that contains one double bond in its carbon chain.
Mousse	A soft or creamy dessert made light by the addition of whipped cream, egg whites, or both.
Muffin method	A mixing method in which the mixed dry ingredients are combined with the mixed liquid ingredients.
Napoleon	A dessert made of layers of puff pastry filled with pastry cream.
Natural sour	See Sourdough starter.
Natural starter	See Sourdough starter.
Net weight	The weight of the total contents of a can or package.
non-fat milk	Milk with all the fat removed.
Non-fat dried milk	Fat-free milk with all the moisture removed.
No-time dough	A bread dough made with a large quantity of yeast and given no fermentation time, except for a short rest after mixing.
Nougatine	A mixture of caramelized sugar and almonds or other nuts, used in decorative work and as a confection and flavoring.
Nouvelle cuisine	Important cooking style of the 1960s and 1970s, known for lighter flavours and elaborate plating styles.
Nutrient	A substance essential for the functioning or growth of an organism.
Nutrient density	The quantity of nutrients per calorie.
Oil	A liquid fat.
Old dough	A dough that is over fermented.
One-stage method	(1) A cookie-mixing method in which all ingredients are added to the bowl at once. (2) A cake-mixing method in which all the ingredients, including high-ratio liquid shortening, are mixed together at once.
Opera cake	A layer cake made of thin sponge layers, coffee-flavoured buttercream, and chocolate ganache.
Opson	In ancient Greece, any food eaten with bread.
Osmotolerant yeast	A type of yeast that can remain active even in a high concentration of sugar. Used for sweet doughs.
Othello	A small (single-portion size), spherical sponge cake filled with cream and iced with fondant.
Outlining	Drawing shapes on a plate with chocolate or a thick sauce prior to filling in the spaces in those shapes with one or more sauces.

Oven spring	The rapid rise of yeast goods in the oven due to the production and expansion of trapped gases caused by the oven heat.
Overrun	The increase in volume of ice cream or frozen desserts caused by the incorporation of air while freezing.
Ovo-vegetarian	Referring to a vegetarian diet that allows eggs.
Oxidation	The process that occurs when oxygen reacts with other compounds or elements. In the bakeshop, it usually refers to oxidation of components of flour during mixing.
Pain d'épice (pan day peece)	French for "spice bread." A type of gingerbread.
Pain de campagne	French country-style bread.
Palmier (palm yay)	A small pastry or petit four sec made of rolled, sugared puff pastry cut into slices and baked.
Panettone	An Italian sweet bread made in a large loaf, generally containing dried and candied fruits.
Panna cotta	An Italian pudding made of cream, gelatine, and flavourings; literally, "cooked cream."
Paper cone	A tool made of parchment paper formed into a cone and filled with icing, sauce, or other semi-liquid. Used for decorative work.
Parasite	An organism that can survive only by living on, with, or inside another organism.
Parfait	(1) A type of sundae served in a tall, thin glass. (2) A still-frozen dessert made of egg yolks, syrup, and heavy cream.
Paris-Brest	A dessert consisting of a ring of baked éclair paste filled with cream.
Pasteurize	To heat-treat substances, such as milk, to kill bacteria that might cause disease or spoilage.
Pastillage (pahs tee yahzh)	A sugar paste used for decorative work, which becomes very hard when dry.
Pastry cream	A thick custard sauce containing eggs and starch.
Pastry flour	A weak flour used for pastries and cookies.
Pastry method	A mixing method in which the fat is mixed with the dry ingredients before the liquid ingredients are added. Also called Biscuit method.
Pâte à choux (paht ah shoo)	Éclair paste.

Pâte brisée	A type of rich pastry dough used primarily for tarts.
Pâte fermentée	Fermented dough, used as a starter.
Pâte feuilleté (PAHT fuh yuh TAY)	French name for puff pastry.
Pâte sablée	A rich, crumbly pastry dough high in fat. Also called short dough.
Pâte sucrée	A type of pastry dough similar to pâte brisée but higher in sugar.
Patent flour	A fine grade of wheat flour milled from the inner portions of the kernel.
Pathogen	A disease-causing microorganism.
Pâtissier (pah tis YAY)	A pastry chef.
Peasant tart	A baked tart with a custard filling containing prunes.
Pectin	A soluble plant fiber, used primarily as a jellying agent for fruit preserves and jams.
Peel	A flat wooden shovel used to place hearth breads in an oven and to remove them.
Pentosan	A category of carbohydrate gums present in wheat flour and having strong water absorption capability.
Petit four (p'tee foor)	A delicate cake, pastry, cookie, or confection small enough to be eaten in one or two bites.
Petit four glacé	An iced or cream-filled petit four.
Petit four sec	An un-iced or unfilled petit four (sec means "dry"), such as a small butter cookie or palmier.
pH	A measure of the acidity or alkalinity of a substance.
Philadelphia-style ice cream	Ice cream containing no eggs.
Phyllo (FEE lo) dough	A paper-thin dough or pastry used to make strudels and various Middle Eastern and Greek desserts.
Pickup stage	The first stage of yeast dough mixing, in which the loose dry ingredients are gradually picked up and incorporated into the developing dough.
Piping jelly	A transparent, sweet jelly used for decorating cakes.
Pithiviers (pee tee vyay)	A cake made of puff pastry filled with almond cream.
Plant toxin	Any poison naturally present in plants.
Poissonier	The station chef in charge of fish preparation.
Polyunsaturated fat	A type of fat, normally liquid at room temperature, that contains more than one double bond in its carbon chain.

Poolish	A thin yeast starter made with equal parts flour and water, plus commercial yeast.
Pot de crème (poh duh krem)	A rich baked custard. Pound cake (1) A cake made of equal weights of flour, butter, sugar, and eggs; (2) any cake resembling this.
Pour batter	A batter liquid enough to pour.
Poured sugar	Sugar boiled to the hard-crack stage and then poured into moulds to harden. Also called cast sugar.
Praline	A confection or flavouring made of nuts and caramelized sugar.
Pre-ferment	A fermented dough or batter used to provide leavening for a larger batch of dough.
Press	A scaled piece of dough divided into small, equal units in a dough divider.
Profiterole	A small puff made of éclair paste. Often filled with ice cream and served with chocolate sauce.
Proofing	Fermenting made-up yeast products to increase their volume and lighten their texture before baking.
Protein	Any of a group of nutrients essential for growth, building body tissue, and basic body functions, and that can also be used for energy if the diet does not contain enough carbohydrates and fats.
Puff pastry	A very light, flaky pastry made from a rolled-in dough and leavened by steam.
Pulled sugar	Sugar boiled to the hard-crack stage, allowed to hardens lightly, then pulled or stretched until it develops a pearly sheen.
Pullman loaf	A long, rectangular loaf of bread.
Pumpernickel flour	A coarse, flaky meal made from whole rye grains.
Punching	A method of expelling gases from fermented dough.
Purée	A food made into a smooth pulp, usually by being ground or forced through a sieve.
Quenelle (kuh NELL)	A small oval portion of food.
Recipe	A set of instructions, including ingredients and quantities, for preparing a certain dish. See also Formula.
Regular shortening	Any basic shortening without emulsifiers, used for creaming methods and for icings.

Retarder-proofer	An automated, timer-controlled combination of retarder/freezer and proofer, used for holding and proofing yeast products.
Retarding	Refrigerating a yeast dough to slow its fermentation.
Reversed puff pastry	A type of puff pastry made with the dough enclosed between layers of butter.
Ribbon sponge	A thin sponge cake layer with a decorative design made of stencil paste.
Rice condé	A thick, moulded rice pudding, usually topped with fruit.
Rice impératrice	A rich rice pudding containing whipped cream, candied fruits, and gelatine.
Rich dough	A dough high in fat, sugar, and/or eggs.
Ripe (fruit)	Fruit that is at its peak of texture, flavour, and sweetness, and ready to be consumed.
Rolled	A cookie makeup method in which the dough is rolled out into a sheet and cut into shapes with cookie cutters.
Rolled fondant	A dough-like sugar product with the texture of a stiff dough. Rolled into thin sheets and used to cover cakes.
Rolled-in dough	Dough in which a fat has been incorporated in many layers using a rolling and folding procedure.
Roller milling	A process of milling wheat into flour that involves repeatedly cracking and sifting the grain.
Rôtisseur (ro tee sur)	Roast cook or meat cook.
Rounding	A method of moulding a piece of dough into a round ball with a smooth surface or skin.
Royal icing	A form of icing made of confectioners' sugar and egg whites; used for decorating.
Rye blend	A mixture of rye flour and hard wheat flour.
Rye flour	Rye grain milled into a flour.
Rye meal	Coarse rye flour.
Sabayon	A foamy dessert or sauce made of egg yolks whipped with wine or liqueur.
Sablage	See Sanding method.
Sachertorte	A rich chocolate cake from Vienna.
Sacristain (sak ree stan)	A small pastry made of a twisted strip of puff paste coated with nuts and sugar.

Sanding method	A pastry- and cookie-mixing method involving blending the fat with the dry ingredients and then adding in egg.
Saturated fat	A fat that is normally solid at room temperature.
Saucier	The station chef who prepares sauces and stews and sautés foods to order.
Savarin	A type of yeast bread or cake soaked in syrup.
Scaling	Weighing, usually ingredients, doughs, or batters. Scone : A type of biscuit or biscuit like bread.
Scone flour	A mixture of flour and baking powder used when very small quantities of baking powder are needed.
Seeding	A technique for tempering chocolate by adding grated tempered chocolate to melted chocolate to cool it.
Self-rising flour	White flour to which baking powder and, sometimes, salt has been added.
Sfogliatelle (SFO lee ah TELL eh)	A southern Italian flaky turnover pastry with a sweet cheese filling.
Sheet	A cookie makeup method in which the dough is baked in sheets and then cut into portions.
Sherbet	A frozen dessert made of water, sugar, fruit juice, and, sometimes, milk or cream.
Short	Having a high fat content, which makes the product (such as a cookie or pastry) very crumbly and tender.
Shortbread	A crisp cookie made of butter, sugar, and flour.
Short dough	A pastry dough, similar to a basic cookie dough, made of flour, sugar, and fat. See also Short.
Shortening	(1) Any fat used in baking to tenderize the product by shortening gluten strands. (2) A white, tasteless, solid fat formulated for baking or deep frying.
Short-fermentation straight dough	A yeast dough, usually with a high percentage of yeast, that is fermented for only a short time before being made up and baked.
Short mix	A yeast dough mixing technique combining a short mixing period with long fermentation.
Simple fold	One part of the procedure for making Danish and croissant dough, which requires folding the dough in thirds. Also called threefold.
Simple presentation	A style of plating a dessert consisting of a portion of one dessert plus optional sauces and garnishes.

Simple sugar	A sugar with the simplest or smallest possible molecule, containing 6 carbon atoms. Glucose is a simple sugar.
Simple syrup	A syrup consisting of sucrose and water in varying proportions.
Single-acting baking powder	Baking powder that releases gases as soon as it is mixed with water.
Skim milk	Milk with all the fat removed.
Sodium bicarbonate	Baking soda; a chemical that releases carbon dioxide gas when combined with an acid.
Soft meringue	The type of meringue traditionally used for pie toppings; usually with a low percentage of sugar.
Soft pie	A single-crust pie with a custard-type filling—that is, a filling that sets or coagulates due to its egg content.
Soft wheat	Wheat low in protein.
Solid pack	A type of canned fruit or vegetable with no water added.
Sorbet (sor BAY)	French for "sherbet."
Sorbetto	Italian for "sherbet."
Soufflé	(1) A baked dish containing whipped egg whites, which cause the dish to rise during baking. (2) A still-frozen dessert made in a soufflé dish so it resembles a baked soufflé.
Sour	Sourdough starter. Sour cream : A cream, usually with about 18% milk fat, that has been fermented by bacteria until thick.
Sourdough	A dough leavened by a sourdough starter.
Sourdough starter	A dough or batter that contains wild yeasts and bacteria, has a noticeable acidity as a result of fermentation by these organisms, and is used to leaven other doughs.
Sous chef	The station chef who assists the executive chef and is directly in charge of the cooking during production.
Sponge	A batter or dough of yeast, flour, and water that is allowed to ferment and is then mixed with more flour and other ingredients to make a bread dough.
Sponge cake	A type of cake made by whipping eggs and sugar to a foam, then folding in flour.
Sponge method	A cake- and cookie-mixing method based on whipped eggs and sugar.

Sponge roll	Swiss roll.
Spread	The tendency of a cookie to expand and flatten out when baked.
Spun sugar	Boiled sugar made into long, thin threads by dipping wires into the sugar syrup and waving them so the sugar falls off in fine streams.
Staling	The change in texture and aroma of baked goods due to the loss of moisture by the starch granules.
Standardized formula	A set of instructions describing the way a particular establishment prepares a certain baked item.
Starch retrogradation	A chemical change of starch molecules that is responsible for staling.
Stencil	A pattern or design cut from plastic or cardboard, used for depositing batter for thin cookies made in decorative shapes.
Stencil paste	A type of thin cookie or wafer dough used to make cookies in decorative shapes; also used to make decorative patterns in ribbon sponge.
St-Honoré	(1) A dessert made of a ring of cream puffs set on a short dough base and filled with a type of pastry cream; (2) the cream used to fill this dessert, made of pastry cream and whipped egg whites.
Stirred custard	A custard stirred while it is cooked so it thickens but does not set.
Stollen	A type of sweet yeast bread with fruit.
Straight dough method	A mixing method for yeast goods in which all ingredients are mixed together at once.
Straight flour	Flour made from the entire wheat kernel minus the bran and germ.
Stream	The portion of flour that is separated in any one of the stages in the roller-milling of grain.
Streusel (STROY sel)	A crumbly topping for baked goods, consisting of fat, sugar, and flour rubbed together.
String work	The production of decorative icing designs using a paper cone; see drop-string method.
Strong flour	Flour with a high protein content.
Strudel	A baked item consisting of a filling rolled up in a sheet of strudel dough or phyllo dough.
Strudel dough	A type of dough that is stretched until paper-thin.
Sucralose	A type of synthetic sweetener derived from sucrose.

Sucrose	The chemical name for regular granulated sugar and confectioners' sugar.
Sugar cage	A lacy dome of hard or caramelized sugar.
Sundae	A dessert consisting of ice cream in a dish, with various sauces and toppings.
Swiss meringue	Egg whites and sugar warmed, usually over hot water, and then whipped to a foam.
Swiss roll	A thin sponge cake layer spread with a filling and rolled up.
Syrup	One or more types of sugar dissolved in water, often with small amounts of other compounds or impurities that give the syrup flavour.
Syrup pack	A type of canned fruit containing sugar syrup.
Tablage	A technique for tempering chocolate by cooling it on a marble slab.
Tart	A flat, baked item consisting of a pastry and a sweet or savoury topping or filling; similar to a pie but usually thinner.
Tarte Tatin	An upside-down apple tart.
Tempering	The process of melting and cooling chocolate to a specific temperature to prepare it for dipping, coating, or moulding.
Tenacity	The resistance of a dough to being stretched.
Three-fold	A technique used to increase the number of layers in puff pastry or Danish pastry by folding the dough in thirds.
Tiramisù	An Italian dessert made of ladyfinger sponge, flavoured with espresso coffee and a creamy cheese filling.
Torte	German for various types of cakes, usually layer cakes.
Trans fat	A solid fat, usually manufactured by hydrogenation, that limits the body's ability to rid itself of cholesterol.
Tulipe	A thin, crisp cookie moulded into a cup shape.
Tunneling	A condition of muffin products characterized by large, elongated holes; caused by overmixing.
Turntable	A pedestal with a flat, rotating top, used for holding cakes while they are being decorated.
Two-stage method	A cake-mixing method that begins with the blending of flour and high-ratio shortening and is followed by the addition of liquids. Also called the high-ratio method.

UHT pasteurization	Ultra Heat Treated or Ultra High Temperature pasteurization. Subjected to a relatively high heat for a defined period in order to kill microorganisms and extend shelf life.
Ultrapasteurized	See UHT pasteurization.
Unsaturated fat	A fat that is normally liquid at room temperature.
Vacherin (vahsh er ran)	A crisp meringue shell filled with cream, fruits, or other items.
Vegan	A vegetarian diet that excludes all animal products, including dairy products and eggs.
Virus	An extremely small microorganism, smaller than bacteria, responsible for a variety of diseases.
Vital wheat gluten	Wheat gluten in a concentrated form, usually about 75% by weight, added to flour to improve the quality of yeast-raised doughs.
Vitamin	Any of a group of compounds that are present in foods in very small quantities and are necessary for regulating body functions.
Wash	(1) n. A liquid brushed onto the surface of a product, usually before baking. (2) v. To apply such a liquid.
Water hardness	The mineral content of water.
Water pack	A type of canned fruit or vegetable containing the water used to process the item.
Weak flour	Flour with a low protein content.
Whipping cream	Cream with a high enough fat content, usually above 30%, to enable it to be whipped into a foam.
White couverture	A confection consisting of cocoa butter, milk solids, and sugar. Sometimes erroneously called white chocolate.
Whole milk	Milk with 3.5% fat content.
Whole wheat flour	Flour made by grinding the entire wheat kernel, including the bran and germ.
Yeast pre-ferment	A dough or batter fermented with commercial yeast and used to provide leavening for a larger batch of dough.
Yeast starter	Another name for yeast pre-ferment.
Yogurt	Milk cultured with special bacteria until thick.
Young dough	A dough that is under fermented.
Zabaglione	An Italian dessert or sauce made of whipped egg yolks and Marsala wine.
Zest	The coloured outer portion of the peel of citrus fruits.

Bibliography

1. Food Safety Management System for Bakery & Bakery Products.
   Food Safety Standard Authority of India
2. Manual for Small Scale Bakery Units
   Food Safety Standard Authority of India
3. Professional Baking
   Fifth Edition (Wayne Gisslen)
4. Professional Baking
   Sixth Edition (Wayne Gisslen)
5. The Art and Science of Culinary Preparation
   Jerald W Chesser
6. Photograph from Google

Index

A

Acesulfame potassium 178
Acetate 156
Acetic acid bacteria 34
Aciduric bacteria 35
Agar-agar 201
Agave nectar 180
Air and dust 43
Air handling unit 27
Algae 38
Allergen control and management 22
Alternaria 37
Amylopectin starches 201
Amylose 201
Amylose and amylopectin starches 165
Animal feeds 44
Antimicrobial substances 40
Approvals required 8
Aroma 239
Artificial sweeteners 177
Aspartame 178
Aspergillus 37

B

Baba mould 146
Bacteria 33
Baker's formulas & recipes 104
Baker's percentages 113
Baker's yeast 222
Bakery 1
Baking 26, 249
Baking ammonia 195
Baking powders 195
Baking soda 194
Balancing fat 262
Banneton 147
Barquette 147
Basic pastries 256
Basic units 110
Bench-proofing 231
Beneficial role 42
Berries and fruit 208
Biological structures 40
Biscuits and cookies 307
Blending 256
Blowtorch 149
Boiled icing 303
Bombe mould 147
Bombes 292, 300
Boulanger 2, 312
Bowl knife 150
Bran 160

Bread faults 243
Bread improver 171
Bread Loaf 244
Bread making 219
Bread mixing methods 236
Bric dough 207
Brioche mould 147
Brown rice syrup 179
Brown sugar 174
Brushes 150
Buckwheat 164
Bulk fermentation 229
Butter 197
Buttercreams 285
Buttermilk 188
Butyric acid bacteria 34

C

Cake and baker's margarines 198
Cake faults 262
Cake formula 261
Cake making 253
Cake pans 141
Cakes 301
Candied nuts 209
Candied zest 209
Canned food 211
Caramelized sugar 205
Ceiling 13
Charlotte ring 148
Cheese 208
Chefs and head cooks 7
Chemical contamination 48
Chemical leaveners 194, 213
Chemical spoilage 50
Chiboust 260
Chinois and China cap 151
Chocolate 205, 264
Chocolate bows 277
Chocolate cigarettes 278
Chocolate curls and shavings 279
Chocolate cutouts 275
Chocolate designs 275
Chocolate foam icing and filling 289
Chocolate moulds 149
Chocolate petals 280
Chocolate strips wood grain pattern 277
Chocolate tempering 273
Choux paste 208
Choux pastry 257
Churn-frozen desserts 293
Cleaning 18
Coating chocolate 272
Cocoa butter 184, 270
Cocoa pods 264
Cocoa powder 184
Cocoa processing 265
Cocoa production 183
Coconut sugar 176
Cold holding food 55
Color formation 217
Compost 97
Compressed yeast 222
Conching 269
Confectionary 4, 252
Container for holding waste 19
Continuous mixer 134
Control of operation 19
Convection oven 139
Cookie faults 250

Cookies 210, 246
Cooks 7
Cooling 26, 235, 249
Corn 163
Corn syrup 179
Cornstarch 200
Cornstick pan 143
Cost calculations 117
Couche 156
Coupes 299
Cream cheese 302
Creaming 195
Creaming method 248
Creme diplomate 259
Critical points 24
Cross-contamination 45
Crust 238
Cultured buttermilk 188
Curled cutouts 276
Curled strips 278
Cuts 60
Cutters 151

D

Dairy products 187
Date sugar 176
Deck oven 138
Delayed-salt method 237
Devices and equipment 108
Dhal flours 162
Diastatic malt 172
Divider-rounder 136
Donating unwanted food 98
Door 14
Dough fermentation trough 136
Dough sheeter 136
Dried fruit 209
Dried yeast 222
Dry ingredients 21
Dry whole milk solids 188
Dryers 261
Drying 217, 267
Dutched cocoa 270

E

Edible garnishes 203
Egg selection 187
Eggs 21, 173, 185
Eggs and dairy products 185
Electrical panel 15
Enriching agents 173
Enrobing 27
Environmental conditions 51
Enzymic browning 50
Enzymic spoilage 50
Equation modifications 107
Equations and measurement 104
Equipment 130
Equipment sanitation and safety 130
Evaporated milk 188
Exhaust fan 15
External characteristics 238
Extrinsic factors 39

F

Fats 212
Fecal coliforms 36
Ferment and dough method 237
Fermentation 190, 192, 266
Filo pastry 206
Fire license 8, 9
First aid 60
Fittings 12

Fixed expense 109
Fixings 107
Fixings unit cost 117
Flaky Pastry 257
Flexipan 142
Floor 12
Flour grades 160
Flour-batter method 255
Foaming 195
Foam-type icings 287
Fondant 284
Food additives 202
Food centre 99
Food contamination 42
Food control and monitoring equipment 19
Food costing 104
Food handlers 44
Food handling 51, 53
Food laws 211
Food microbiology 32
Food packaging 28
Food preparation 54
Food preparation workers 8
Food preservatives 202
Food processing 24
Food processing workers 7
Food receipt 19
Food Safety and Standards Authority of India (FSSAI) 8, 9, 12-14, 19, 30, 31, 92-95, 181, 183
Food safety hygiene 32
Food scraps 101
Food spoilage 49
Food traceability 30
Food transportation 29
Food utensils 45
Food waste scheme 98
Forming 26
Frangipane 259
Freezer storage 304
Frozen dessert 292
Fruit combinations 181
Fruit concentrates 176
Fryer 140
FSSAI license 8, 19
FU Dge-type icings 289
Fungi 36

G

Ganache 303
Garbage disposal 100
Gas-producing bacteria 35
Gastrointestinal tract 44
Gelatine 200
Gelling agents 199
Genetically modified foods 215
Genoise method 255
Geotrichum 37
Glazes 291
Gloves 59
Glycerol mono stearate 172
Gram flour 162
Granulated brown sugar 174
Granulated sugar 174
Grinding 268
Grocery store 99
GST registration 8, 9
Gum tragacanth 202

H

HACCP implementation plan for cakes 89

HACCP implementation steps 67
HACCP plan for bread 80
HACCP plan for cakes 85
Halotolerant bacteria 35
Hand washing 52, 59
Harvesting 266
Hazard analysis & identification for biscuits & pies 73
Hazard analysis & identification for bread 81
Hazard analysis & identification for cakes 86
Hazard analysis and critical control point (HACCP) 64, 66
Health status 55
Heavy cream 189
High ratio plastic shortenings 197
Holding waste 19
Homemade ice creams 296
Honey 178
Horizontal mixer 135
Hot holding 55
Hot water crust pastry 258
Hotel pan 143
Hotels confectionary 6
Hydrogen ion concentration 39
Hydrometer 158

I

Ice cream freezer 157
Icing 284
Icing comb 151
Illness & injury 56
Ingredient functions 261
Inspection of raw material 20
Installation, fittings, fixtures 18
Instructional recipe and formulas 106
Internal characteristics 239, 260
Internal structures 12
Intrinsic factors 39
Isomalt 177
Isomalt sugar 205

J

Jam 182

K

Kataifi dough 207

L

Lactic acid bacteria 34
Lapsi fada 162
Lard 199
Large equipment 131
Dough handling 136
Layout 8
Layout design and construction 10
Leavening agents 254
Light 15
Light cream 189
Lipolytic bacteria 34
Loaf pan 144
Local charities 98
Local municipal corporation health license 8
Location and surroundings 9
Low fat milk 188

M

Madeleine pan 143
Maida 161
Main equipments for food safety 19
Maintaining quality 187
Maple sugar 175, 176
Maple syrup 179

Marble top table 157
Margarine 198
Marmalades 183
Marshmallow icing 288
Marzipan/rolled fondant 206
Material of construction 17
Material preparation 24
Measurement 108
Mechanical oven 139
Meringues 204
Mesophilic bacteria 41
Microbes 39
Microbial contamination 42
Microbial food contamination 47
Microbial spoilage 49
Microorganism 32
Milk and milk products 173
Milk chocolate 273
Milk powder 188
Milk products 212
Mineral additives 171
Mixers 131
Mixing 25, 224
Mixing bowls 146
Mixing methods 247
Modified straight dough method 225
Moisteners 261
Moistening agent 254
Moisture content 40
Molasses 175, 179
Molds 37
Moulder 137
Mousseline 260
Muffin pan 142

N

Natural cocoa 269
Non-conforming products 28
Non-diastatic malt 172
Non-reusable food waste 99
No-time dough method 236
Nutrients content 40

O

Oils 199
Oils and fats 100
One-stage method 247
Oraganic food 214
Organic acid 172
Organic eggs 186
Organization chart 7
Osmophilic bacteria 35
Oven spring 217
Oven temperature 216
Ovens 137

P

Packaging 51
Panning 232, 249
Pans, containers, moulds 141
Parasites 38
Parchament paper 158
Parfaits 299
Pastry 5, 206, 303
Pastry bag 152
Pastry ingredients 258
Pastry margarines 198
Patent flour 160
Pectin 201
Peel 152
Penetration wounds 63

Personal behaviour 57
Personal cleanliness 56
Personal hygiene 55
Perusulphates 171
Pests 47
Petit four moulds 149
pH values 39, 166
Phyllo 257
Physical contamination 49
Physical spoilage 49
Pie pan 145
Piping chocolate 281
Plain boiled icing 288
Police eating house license 8
Poor personal hygiene 46
Popular ice cream desserts 295
Potassium bromate 172
Powdered sugar 174
Premixing 25
Processing 24
Procurement of raw material 19
Proofer 137
Proofing 232
Propionic acid bacteria 34
Proteolytic bacteria 34
Psychrophilic microorganisms 41
Psychrotrophic bacteria 35
Puff pastry 207, 257

Q

Quality control 19, 30
Quality grades 187

R

Rack oven 138
Raw material 23
Raw sugar 174
Ready-to-eat food 52
Ready-to-eat products 21
Recycling and Composting Food Waste 97
Recycling centre 97
Refined fructose 176
Regular shortenings 196
Relative humidity 41
Retarder-proofer 137
Rice flour 162
Rice noodles 210
Roasting 267
Rolled coatings 292
Roller docker 153
Rolling pins 153
Rounding 231
Royal icing 290
Rubbing method 256
Rye flour 163

S

Saccharine 178
Saccharolytic bacteria 35
Safe food handling tips 52
Sanitary procedure 51
Sanitation 32
Savarin mould 149
Scaling ingredients 224
Scoring 234
Scrapers 153
Seeding method 275
Septic system 101
Sheet pan 144
Shelf life 306
Shop smart 102
Short crust pastry 257

Short mix 227
Shortened cakes 262
Shortenings 196
Shut wounds 63
Sieve 154
Silicone mat 156
Skim milk 188
Slicing/packing of products 27
Slime producers 35
Sodium chloride 221
Soft chocolate fans 279
Soft chocolate strips 280
Soil, water and plants 44
Sorbet 297
Soy 164
Speciality ingredients 22
Spiral mixer 133, 134
Spoilage 42
Spoiled foods 99
Sponge and dough method 237
Sponge method 226, 248
Sponges 210
Springform pan 141
Stabilizers 199
Standard brown eggs 186
Standard white eggs 185
Standardized recipes 105
Starches 165, 200
Steam 196
Steam-jacketed kettle 140
Storage 23
Straight dough method 225, 236
Strainer 154
Sucralose 177
Sugar 204
Sugar and sweeteners 174
Sugar butter process 255
Sugar confectionery 305
Sugar cooking stages 281
Swachh Bharat Mission 96
Sweetened condensed milk 188
Sweetened dark chocolate 271
Syrups 178, 283

T

Tabling method 274
Tart pan 145
Temperature 40
Thermoduric bacteria 35
Thermometers 158
Thermophilic bacteria 35, 41
Topping 284
Tougheners 261
Tube pan 141
Tuiles 204
Turntable 155

U

Unsafe temperature 45
Unsweetened or bitter chocolate 271
Urad Dhal flour 163

V

Variable cost 109
Vegetable gums 201
Vegetarian eggs 186
Vehicle inspection 20
Vertical mixer 131, 132
Viruses 32
Visitors 58
Vitamin-enhanced eggs 186

W

Wall 13
Washes 234
Waste disposal 96
Water absorption power 169
Water content 51
Waxy maize 200
Wet ingredients 21
Wheat 159
Wheat flour 219
Whipped cream 302
Whisk 155
White chocolate 273
Whole milk 187
Window 14
Winnowing 268
Wire rack 155
Wood-fired oven 138
Wounds 62

Y

Yeast 36, 189
Yeast foods 172
Yogurt 189

COLOUR PLATES

Butter for sale in uncovered containers open to the air. (P. 43)

Recycling and Composting Food Waste (P. 97)

Leavening Agents (P. 189)

Shortenings (P. 196)

Gelling agents and stabilizers (P. 199)

(P. 203)

Tuiles

Meringues (Hard) (P. 204)

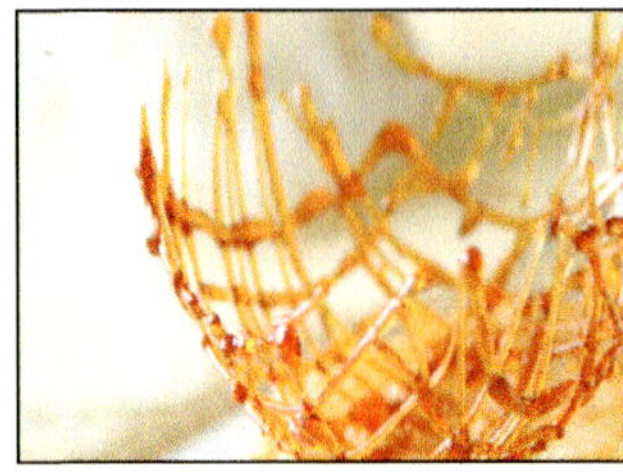

Sugar

Caramelized Sugar (P. 205)

Isomalt Sugar (P. 205)

Chocolate (P. 205)

Marzipan/Rolled Fondant (P. 206)

Filo pastry (P. 206)

Kataifi dough (P. 207)

Bric dough (P. 207)

Puff pastry (P. 207)

Choux paste (P. 208)

Cheese (P. 208)

Berries & Fruit (P. 208)

Dried fruit (P. 209)

Candied zest (P. 209)

Candied Nuts and Brittles (P. 209)

Rice Noodles (P. 210)

Cookies (P. 210)

Sponges (P. 210)

Canned Food (P. 211)

Short Mix (P. 227)

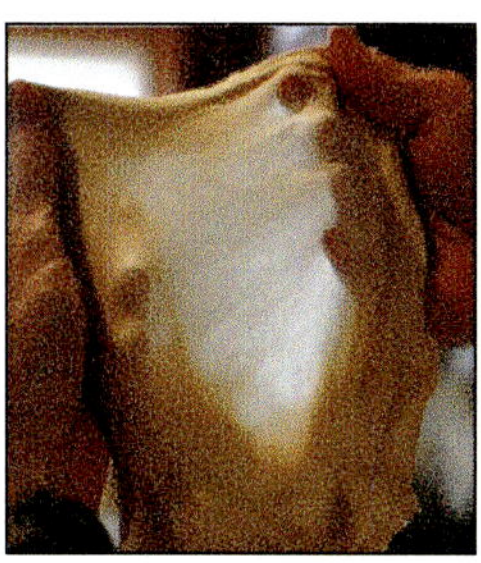

Improved Mix (P. 228)

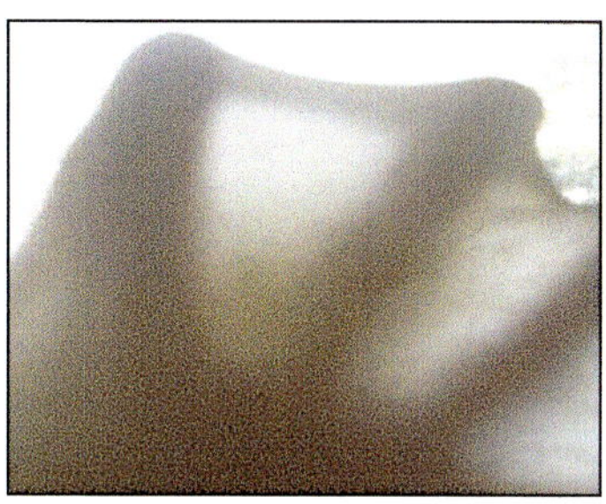

Intensive Mix (P. 228)

Folding (P. 230)

Dividing (Scaling or Portioning Dough) (P. 230)

Bench-Proofing or Intermediate Proofing (P. 231)

Makeup and Panning (P. 232)

Proofing (P. 232)

Washes (P. 234)

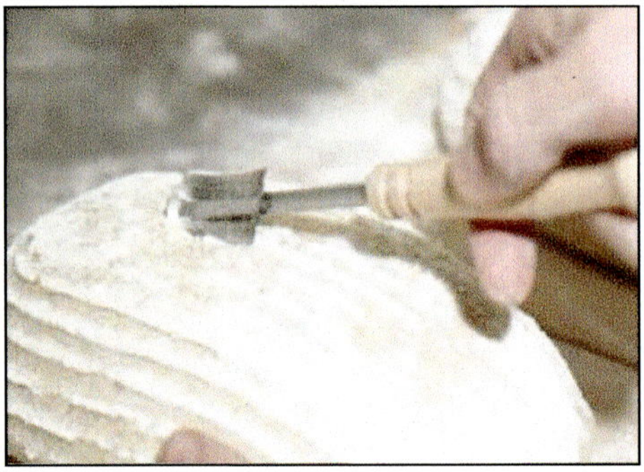

Scoring (P. 234)

Cookies and Biscuits (P. 246)

One-Stage Method (P. 247)

Creaming Method (P. 248)

Sponge Method (P. 248)

Cake (P.252)

Cocoa (P. 264, 266)

Cocoa Fermentation (P. 266)

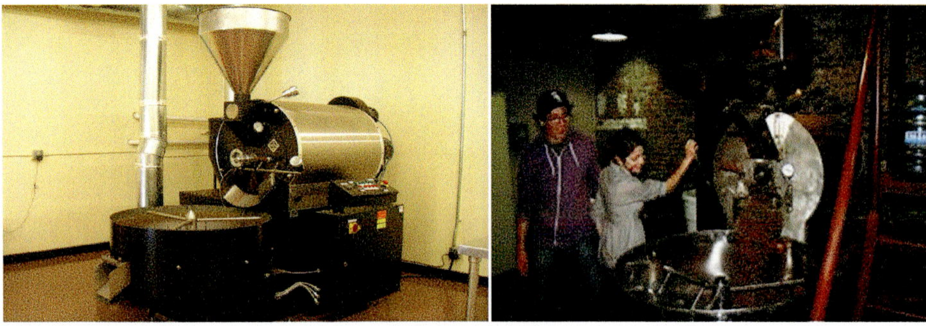

Roasting (P. 267)

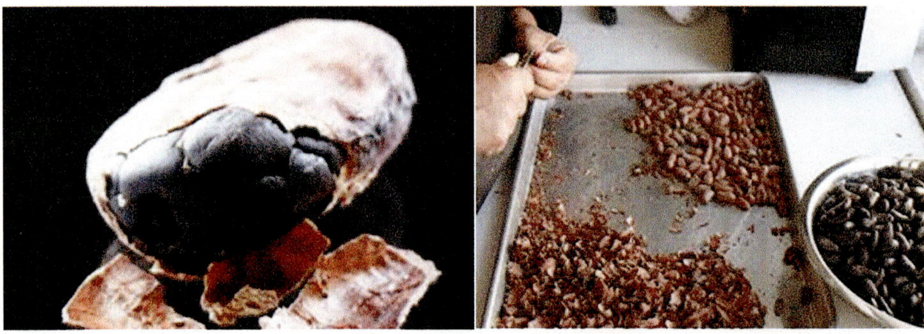

Winnowing (P. 268)

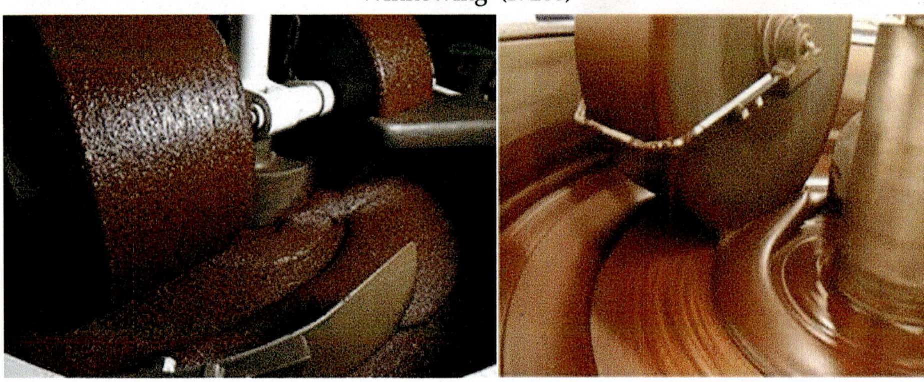

Grinding (P. 268)

Conching (P. 269)

Natural Cocoa (P. 269)

Dutched Cocoa (P. 270)

Cocoa Butter (P. 270)

Bitter Chocolate (P. 271)

White Chocolate (P. 273)

Milk Chocolate (P. 273)

White Chocolate (P. 273)

Dark Chocolate (P. 271)

Chocolate Cutouts (P. 275)

Curled Cutouts (P. 276)

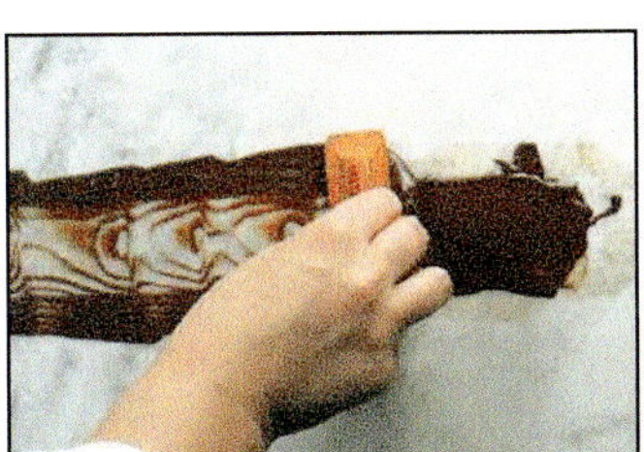

Chocolate Strips Wood Grain Pattern (P. 277)

Chocolate Bows: (P. 277)

Curled Strips (P. 278)

Tabling Method (P. 274)

Chocolate Cigarettes (P. 278)

Chocolate Curls and Shavings (P. 279)

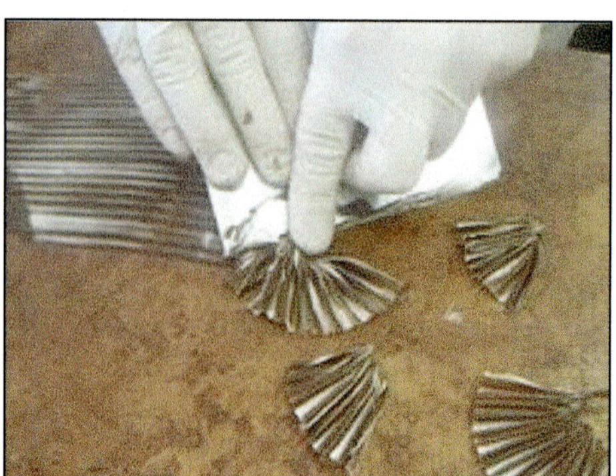

Soft Chocolate Fans (P. 279)

Soft Chocolate Strips (P. 280)

Chocolate Petals (P. 280)

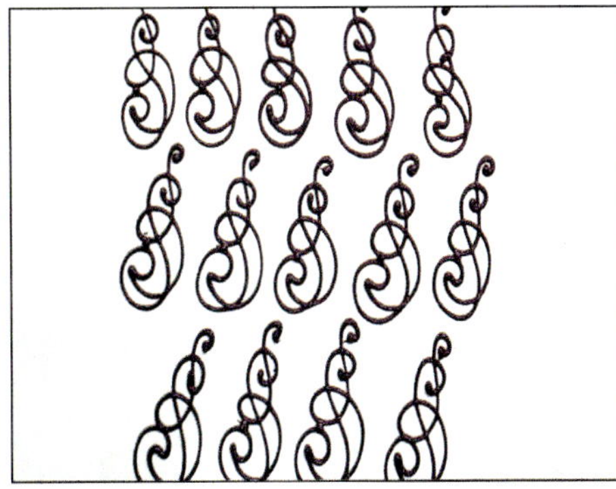

Piping Chocolate (P. 281)

Sugar Cooking Stages (P. 281)

Sugar for Desert and Presentation (P. 283)

Caramelised Sugar Preparation (P. 282)

Icing and Topping (P. 284)

Fondant (P. 284)

Buttercreams (P. 285)

Foam-Type Icings (P. 287)

Chocolate Foam Icing and Filling (P. 289)

Plain Boiled Icing (P. 288)

Marshmallow Icing (P. 288)

FU Dge-Type Icings (P. 289)

Glazes (P. 291)

Royal Icing (P. 290)

Rolled Coatings (P. 292)

Vanila (P. 299)

Parfaits (P. 299)

Bombes (P. 300)

Cassata Napoletana (P. 301)

Enzymic Spoilage (P. 50)

Chemical Contamination of Food (P. 48)

Pests (P. 47)

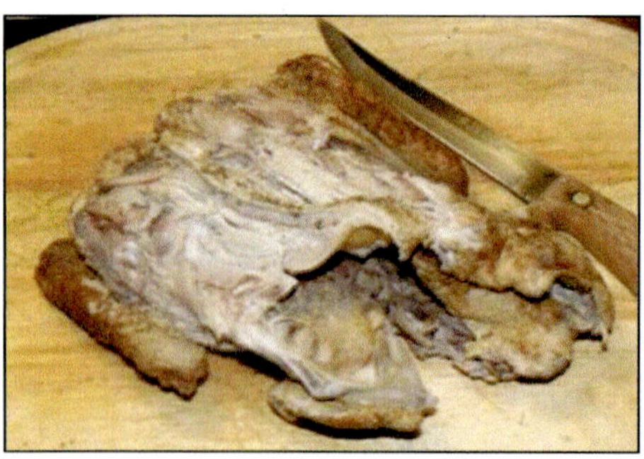

Cross-Contamination (P. 45)

Maintenance and Handling (P. 17)